VI Scientific Technical Conference on "Low-Temperature Plasma During the Deposition of Functional Coatings" 2014

Journal of Physics: Conference Series Volume 567

Kazan, Russia
4-7 November 2014

ISBN: 978-1-63439-673-8
ISSN: 1742-6588

VI Scientific Technical Conference on "Low-Temperature Plasma During the Deposition of Functional Coatings" 2014

Journal of Physics: Conference Series Volume 567

Kazan, Russia
4-7 November 2014

ISBN: 978-1-63439-673-8
ISSN: 1742-6588

IOP Institute of Physics

Printed from e-media with permission by:

Curran Associates, Inc.
57 Morehouse Lane
Red Hook, NY 12571

Some format issues inherent in the e-media version may also appear in this print version.

Printed by Curran Associates, Inc. (2015)

Institute of Physics
Dirac House, Temple Back
Bristol BS1 6BE UK

Phone: 44 1 17 929 7481
Fax: 44 1 17 920 0979

techtracking@iop.org

Additional copies of this publication are available from:

Curran Associates, Inc.
57 Morehouse Lane
Red Hook, NY 12571 USA
Phone: 845-758-0400
Fax: 845-758-2634
Email: curran@proceedings.com
Web: www.proceedings.com

Table of contents

Volume 567

2014

VI Scientific Technical Conference on "Low-temperature plasma during the deposition of functional coatings"
4–7 November 2014, Kazan, Russia

Accepted papers received: 13 November 2014
Published online: 28 November 2014

Preface

Papers

Journal of Physics: Conference Series **567** (2014) 011001 doi:10.1088/1742-6596/567/1/011001

Preface

The VI Republican Scientific Technical Conference "Low-temperature plasma during the deposition of functional coatings" took place from 4 to 7 November 2014 at the Academy of Sciences of the Republic of Tatarstan and the Kazan Federal University. The conference was chaired by a Member of the Academy of Sciences of the Republic of Tatarstan Nail Kashapov – Professor, Doctor of Technical Sciences– a member of the Scientific and Technical Council of the Ministry of Economy of the Republic of Tatarstan.

At the conference, the participants discussed a wide range of issues affecting the theoretical and computational aspects of research problems in the physics and technology of low-temperature plasma. A series of works were devoted to the study of thin films obtained by low-temperature plasma. This year work dedicated to the related field of heat mass transfer in multiphase media and low-temperature plasma was also presented. Of special interest were reports on the exploration of gas discharges with liquid electrolytic electrotrodes and the study of dusty plasmas.

Kashapov Nail, D.Sc., Professor (Kazan Federal University)

Peer review statement

All papers published in this volume of *Journal of Physics: Conference Series* have been peer reviewed through processes administered by the proceedings Editors. Reviews were conducted by expert referees to the professional and scientific standards expected of a proceedings journal published by IOP Publishing.

The influence of the mass flow rate of the electrolyte through the following cathode on the energy characteristics of the gas discharge

Kh K Tazmeev[1], I M Arslanov[1] and G Kh Tazmeev[2]

[1] Branch of Kazan Federal University in Naberezhnye Chelny, 68/19 Mira Street, Naberezhnye Chelny, 423810, Russia
[2] Kazan National Research Technical University named after A.N.Tupolev, 10 K.Marx Street, Kazan,

E-mail: tazmeevh@mail.ru

Abstract. Experimentally investigated the energy characteristics of the gas discharge between the flowing electrolyte cathode and a metal anode at a current 7 – 22 A and capacities 8 to 33 kW. It is found that when changing the mass flow rate of the electrolyte through the flowing cathode, it is possible to adjust the combustion mode of the discharge and to receive a stream of steam plasma with average mass temperature of about 3000 K.

1. Introduction

Gas discharge with liquid electrolyte cathode is of practical interest as a source of nonequilibrium atmospheric pressure plasma. Such a plasma can be promising for the creation of high-temperature vapor-gas medium in the plasma-chemical reactors designed for energy-intensive plasma technologies, in particular for the conversion of hydrocarbons into synthesis gas [1, 2]. For rational use of gas discharge in practice it requires detailed energy characteristics and laws of their change in technologically acceptable intervals power. The aim of this work was to obtain a gas discharge with electrolyte cathode in such capacity and study the influence of the mass flow rate of the electrolyte on its energy characteristics.

2. Experiment

The flowing cathode consisted of a cylindrical vessel *1*, from which flowed the electrolyte *2*. The outer diameter of the vessel *1* was 75 mm and inside it was mounted graphite plate *3* for the supply current. The experimental setup is shown in figure 1.

The discharge was ignited between the liquid electrolyte *2* and the metal anode *4*. The distance l between them was 10 cm. Power supply was served from a three-phase full-wave rectifier connected to the secondary winding step-up transformer. The pulsations of voltage were smoothed with using *C-L-C* filter. The current is changed stepwise variation of the ballast resistor *R*. To measure the electrical parameters used equipment and methods, are described in detail in work [3].

The electrolyte used was a solution of sodium chloride in distilled water with a mass concentration of 5.5 g/l. This concentration of electrolyte required to ensure the sustainability of the discharge at high currents. Due to alkalization of the electrolyte due to the electrolysis process, was made adjustments to its electric conductivity σ is the before beginning each experience. Measurements σ were carried out on conductometer ANION 4150. The duration of the experiments were chosen so that the increase σ shall not exceed 5 percent of its original value, which was 10 mS/cm. To reduce σ was added distilled water.

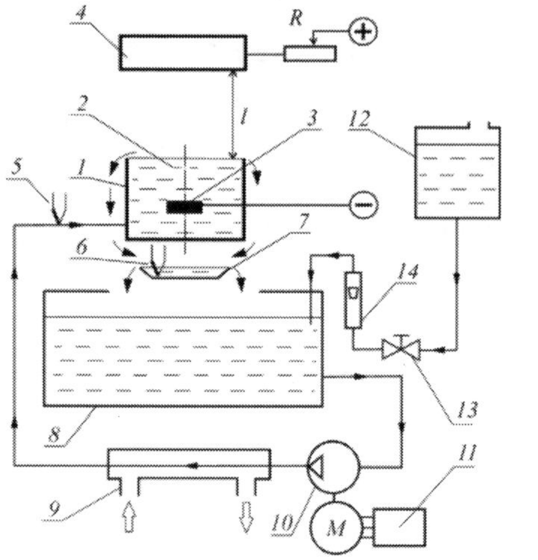

Figure 1. The experimental facility.

Heat Q_k spent on heating of the electrolyte was determined by calorimetric method. The electrolyte temperature was measured chromel-alumel thermocouples *5* and *6*, one of which was mounted at the upstream end of the cylindrical vessel *1* and the other on the small tank *7* located underneath. The electrolyte flowing out of the cylindrical vessel *1* almost fully fall into this small container *7*, and then drained into the reservoir *8*. Removal of heat from the electrolyte was carried out by passing it through a heat exchanger *9*, water-cooled. A frequency converter *11* changed the number of revolutions of the motor M of the hydraulic pump *10*. In this way ensured the continuity of mass flow rate electrolyte *m* through the flowing cathode. Pre experimentally was obtained the dependence m from frequency of rotation via frequency converter.

The amount of electrolyte was carried away from the circulation system during the combustion of the gas discharge, but was offset by the addition of distilled water from the feed tank *12*. The flow rate of valve *13* was set so that the level of electrolyte in the tank *8* was on the same height. The consumption of distilled water was determined according to the testimony of the float flowmeter *14* and it was adopted for the mass flow rate of the gas mixture *G* entering in the discharge region between the cathode and the anode.

3. Results and their discussion

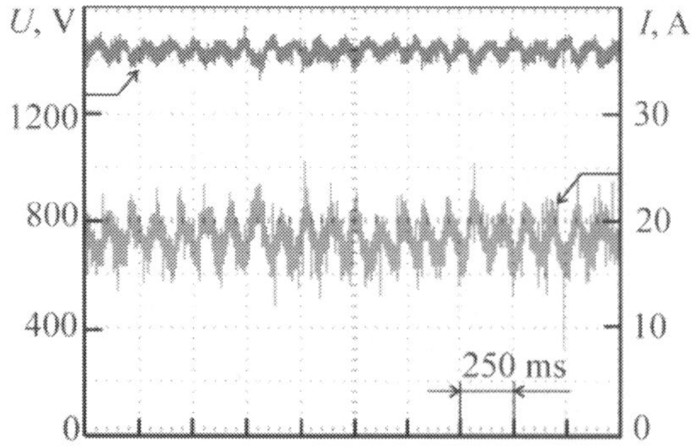

Figure 2. Oscillograms of current and voltage. $m = 20$ g/s; $R = 0$.

Figure *2* shows oscillograms of the discharge current *I* and voltage *U*, which were supplied to the terminals of the cathode and anode. It is seen that both of these parameters are exposed to pulsations. As was established in [4], such low-frequency pulsations with significant amplitudes caused by the change of the shape and size of the plasma column open discharge. Due to the presence of pulsations in the graphical representations of experimental results were used average values of the parameters. In the figures and in the text they were denoted with common signs in the form of angle brackets.

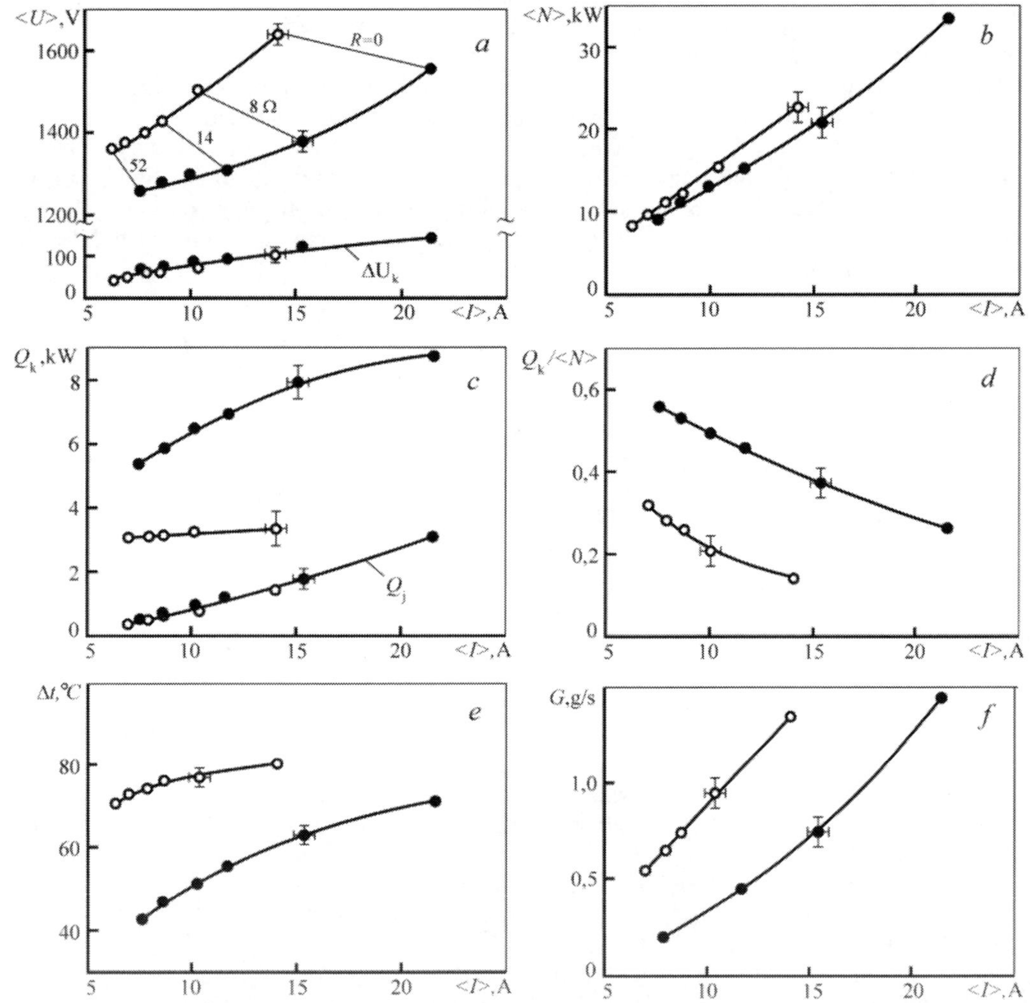

Figure 3. Dependencies on the discharge current, the voltage at the terminals of the electrodes (*a*) power consumption (*b*), the heat losses at the cathode (*c*), the proportion of thermal losses at the cathode in the energy balance (*d*), the increment of temperature of the electrolyte (*e*) and the mass velocity loss of electrolyte (*f*). m = 10 g/s (bright circles) and m = 30 g/s (dark circles).

The voltage measured at the terminals of the cathode and the anode consists of a voltage drop in the electrolyte ΔU_k and voltage in the discharge gap. The values of the ΔU_k, measured with different *m*, practically coincided (figure 3*a*). Therefore, from the comparison of the graphs of the dependence of <*U*> and <*I*> clearly showed that reducing the mass flow of the electrolyte leads to an increase of the voltage in the discharge gap. The total power consumed in the discharge gap and the inside of the electrolyte (Joule heat) increases slightly (figure 3*b*). The change in m has virtually no effect on the power Joule heat dissipation Q_j inside of the electrolyte (figure 3*c*), and thermal losses at the cathode Q_k significantly reduced with decreasing *m*. This means that at a small mass flow of electrolyte takes less heat from the discharge zone. The cost of heating of the electrolyte reduces and their share in the energy balance decreases (figure 3*d*). Previously, this pattern was identified in [5], which studied gas discharge generated under other conditions differ substantially from conditions in this work. It is noteworthy that with decreasing *m*, the proportion of heat losses in the energy balance becomes almost the same as the arc water-steam plasmatron [6].

As expected, for small m, the electrolyte is heated to higher temperatures (figure 3*e*). This significantly increases the amount of electrolyte carried away from the circulation system, ie, increased mass flow rate of electrolyte loss G (figure 3*f*). Basically, the decline is due to the electrolyte spraying and evaporation under the influence of the gas discharge in the area of its binding to the cathode. Therefore, the value G may be considered as a stream of plasma-forming substance. In this embodiment, to simplify the enthalpy of the medium in the plasma column above the electrolyte can be calculated from the formula:

$$i = (< N > - Q_k) / G.$$

Because the electrolyte is weakly concentrated (mass fraction ~ 0.055%), we can assume that the as plasma-forming substance is used mainly water vapor and calculation results i can be attributed to the water vapor plasma. Comparison of calculated values of i with data presented in [7] for steam plasma, allowed us to estimate the mass-average temperature in the plasma column between the cathode and the anode. Its maximum value was within 3000-3200 K.

4. Conclusions

Reducing the mass flow of the electrolyte through the flowing cathode can reduce heat loss and increase the efficiency of electrical energy to create a plasma.

References

[1] Artemov A V, Pereslavtsev A V, Krutyakov Y A, Voschinin C A, Kudrinsky A A, Bulba V A and Ostry I I 2011 *Ecologia i promeshlennost Rossii* **9** 20
[2] Fridland S V, Tazmeev A Kh, Miftakhov M N and Tazmeev Kh K 2006 *Vestnik mashinostroeniya* **7** 72
[3] Tazmeev Kh K, Arslanov I M and Tazmeev G Kh 2014 *Izvestija VUZov. Fizika* **57** *№3-2* 227
[4] Tazmeev Kh K, Arslanov I M and Tazmeev G Kh 2013 *Vestnik KGTU im. A N Tupoleva* **69** *№2-2* 166
[5] Tazmeev Kh K and Tazmeev A Kh 2005 *Heat Transfer Research* **36** 623
[6] Anshakov A S, Rad'ko S I, Urbakh E K and Cherednichenko V S 2014 *Izvestija VUZov. Fizika* **57** *№3-2* 40
[7] Dautov G Y, Timoshevskii A N, Urjukov B A et al 2004 *Generacija nizkotemperaturnoj plazmy i plazmennye tehnologii: Problemy i perspektivy* (Novosibirsk: Nauka)

Journal of Physics: Conference Series **567** (2014) 012002 doi:10.1088/1742-6596/567/1/012002

Apparatus and method for determining the gas permeability and flux of helium through the materials and coatings

V T Barchenko[1], **A A Lisenkov**[1], **M L Vinogradov**[2]
[1] Institute of Problems of Mechanical Engineering, Saint-Petersburg, Russia
[2] Saint Petersburg State Electrotechnical University "LETI" (ETU), Saint-Petersburg, Russia

E-mail: mv@vactron.org

Abstract. Apparatus and method for measuring flow of helium through the materials and coatings, obtained by ion-plasma technologies, are developed and tested. The apparatus for the measurement is designed on the basis of a helium leak detector TI1-14, produced by JSC "Zavod Izmeritel", that provides a minimum detectable flow of helium 7.10-13 Pa.m3/s. The purpose of the study is the creating apparatus and method to determine gas permeability and helium flux through new materials and coatings to create the hermetic devices with special properties. This devices are made from polymer coated with metals, and they should replace full metals device analogues in the field of aerospace engineering.

1. Introduction

The study of gas permeability trough new materials and coatings is important to create the devices of these materials. It is necessary to determine polymer type to create a sealed devices which has minimized gas permeation during operation. Helium is selected as a tracer gas in such tests, because of the high value of its diffusion coefficient through the solid body relative to other gases.

Apparatus and method for measuring flow of helium through the materials and coatings, obtained by ion-plasma technologies, are developed and tested in the laboratory of the electronic instruments and devices department Petersburg State Electrotechnical University.

2. Theory of the gas permeability through the materials and coatings

Permeation is commonly viewed as a three-step process. First, gas dissolves into the solid's surface; then it diffuses through the solid; and finally, it desorbs from it. The time dependence of the permeation rate is determined by the diffusion process because the solution and desorption occur on much shorter timescales than the diffusion through the solid [6]. The concentration c of the diffusing species in the solid's surface is given by Henry's law

$$c = S \cdot p, \tag{1}$$

where S is the solubility coefficient and p is the gas pressure. Diffusion of the dissolved gas into the interior is governed by Fick's second law,

$$\frac{\partial c}{\partial t} = D \cdot \nabla^2 c, \tag{2}$$

where D is the concentration-independent, isotropic diffusion coefficient. If we consider a planar membrane with a large area-to-thickness ratio, equation (2) can be reduced to a one-dimensional form

$$\frac{\partial c}{\partial t} = D \cdot \left(\frac{d^2c}{dx^2}\right), \tag{3}$$

where x is the coordinate perpendicular to the membrane. Assuming that the membrane is initially free of gas and that the concentrations of gas at the two surfaces are fixed at c_l and c_2, respectively, the initial and boundary conditions are

$$c(x,\ t=0) = 0 \tag{4a}$$
$$c(x=0,\ t) = 0 \tag{4b}$$
$$c(x=d, t) = c_2 \tag{4c}$$

where d denotes the membrane thickness. The solution of equation (3) consistent with equations (4a), (4b), and (4c) is [7]

$$c(x,t) = c_1 + (c_2 - c_1)\frac{x}{d} + \frac{2}{\pi}\sum_{n=1}^{\infty} \frac{c_2 \cdot \cos(n\pi) - c_1}{n} \sin\left(\frac{n\pi x}{d}\right) \exp\left(-\frac{Dn^2\pi^2 t}{d^2}\right). \tag{5}$$

The flux of diffusing species is given by Fick's first law,

$$Q = -D\nabla c. \tag{6}$$

The gas flux through the low-pressure surface of the wall can then be calculated from equations (5) and (6):

$$Q(x=d,t) = \frac{D(c_2 - c_1)}{d} + \frac{2D}{\pi}\sum_{n=1}^{\infty} [(-1)^n c_2 - c_1] \exp\left(-\frac{Dn^2\pi^2 t}{d^2}\right). \tag{7}$$

For large values of t one obtains the steady state flux

$$Q(x=d, t\to\infty) = \frac{D(p_2 - p_1)}{d}. \tag{8}$$

The product

$$K = DS \tag{9}$$

is the permeation coefficient. It has, like the diffusion and solubility coefficient, a strong dependence on temperature. Finally, the steady state permeation flux Q of a gas exposed with a partial pressure difference $(p_1 - p_2)$ to a membrane of thickness d and surface area A is

$$Q = A \cdot Q(x=d, t\to\infty) = K\frac{A(p_2 - p_1)}{d}. \tag{10}$$

3. The apparatus for the gas permeability measurement

The apparatus for the measurement is designed on the basis of a helium leak detector, which is implemented, in accordance with OST 5.0170-81 [1], the method of the helium chamber. The helium mass spectrometer leak detector TI1-14, produced by JSC "Zavod Izmeritel" (Saint-Petersburg), provides a minimum detectable flow of helium $7 \cdot 10^{-13}$ $Pa \cdot m^3/s$ [5], which corresponds to the first class of tightness according PNAEG-7-019-89 [2].

The measurement stand is equipped with a vacuum testing chamber for the polymer object installation, its pre-drainage and creating a pressure gradient of helium relative to the outer and inner side of the test object. The vacuum circuit of the measurement stand is shown in Figure 1.

Test object, placed in a vacuum testing chamber, is connected to the leak detector and evacuated from the outside and the inside with a leak detector vacuum system. Helium is supplied into the chamber with continuous pumping of the opposite side of the object.

Quantitative value of the helium flux is measured by leak detector signal after stabilization of the helium diffusion through the test object. The appearance of the stand for gas permeability measurement of materials and products on helium is shown in Figure 2. 3D-model of vacuum chamber for installation the test object is shown in Figure 3. The verification of the samples parameters shall be carried out under normal climatic conditions unless otherwise specified.

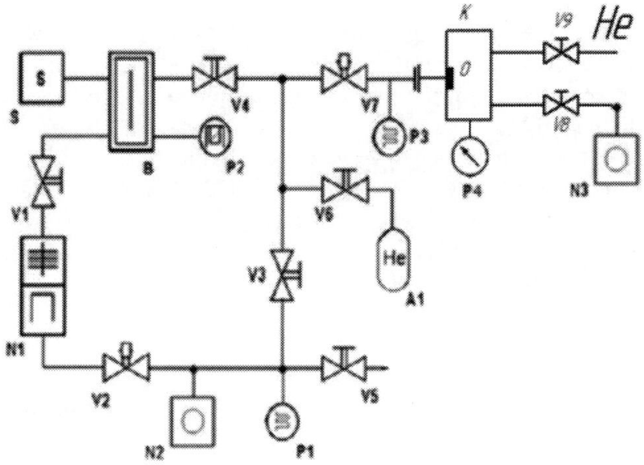

where S – the mass spectrometry analyzer; V1, V3, V4, V6, V8, V9 – manual valve; N1 – turbomolecular pump; V2, V5, V7 – solenoid valve; N2, N3 - rotary vane pump; A1 – helium standard leak; B – cold trap; K – vacuum chamber; O – test object; P1, P3 – Pirani pressure sensor; P2 – ionization pressure sensor; P4 – mechanical manometer.

Figure 1. Vacuum circuit of the apparatus for measuring the flux of helium through the materials and coatings.

Figure 2. The appearance of the stand for measuring the flux of helium through the materials and coatings.

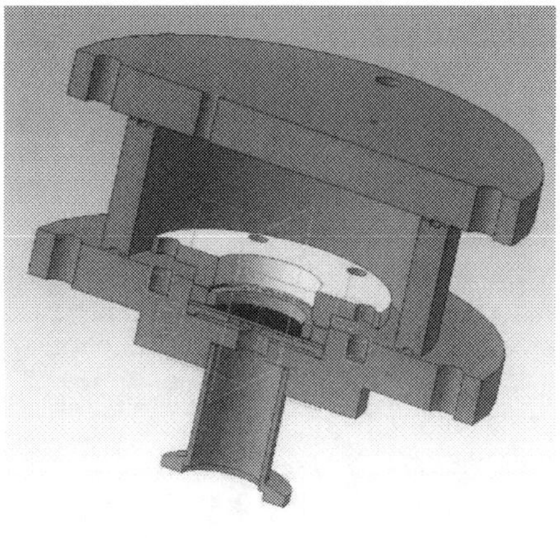

Figure 3. 3D-model of vacuum chamber for installation the test object.

Samples for the measurement should have a disk shape with the following dimensions:

- diameter 44.0 $_{-0.2}$ mm;

- thickness of 1,00 ± 0,05 to 8,00 ± 0,05 mm;
- surface roughness is not less than Ra = 2.5.

According to GOST 28517-90 [3], the test conditions (pressure drop, the direction of the gas load and others) should be established identical as operation conditions of the material or the device. Test conditions other than the operating conditions may be given in a technically and economically justified cases.

4. Testing procedure

The authors developed a technique of testing to determine the flux of helium through the material. Methods of measurement are made according to OST 11-0808-92 [5]. The key points of testing procedure allows to describe in detail the principles of measurements and the apparatus operation.
1 Preparation of the vacuum chamber.
1.1 Clean the inner surface of the chamber, using calico and ethyl alcohol.
1.2 Dry the chamber for 2 minutes.
2 Preparation of the material sample.
2.1 Measure the wall thickness of the sample *d*.
2.2 Install the sample in position for measurement in vacuum chamber.
2.3 Fix the presser cover bolts.
3 Setting up the leak detector.
3.1 Turn on the leak detector in accordance with the instruction manual [4].
3.2 Perform the leak detector preparation to work according to the instruction manual [4].
4 The assembling of the vacuum system.
4.1 Attach the hollow cylinder and the cap of the vacuum chamber to the inlet flange of the leak detector with mounting bolts.
5 Determination of the system sensitivity threshold.
5.1 Determine the sensitivity threshold Q_{min} by a test helium leak, according PNAEG-7-019-89 [2].

The sensitivity threshold of the helium chamber measurement method depends on the magnitude of the signal background oscillations $\Delta\alpha_f$ and is given by:

$$Q_{min}=2\Delta\alpha_f \cdot S_Q, \tag{11}$$

where $\Delta\alpha_f$ is the maximum amplitude of the background signal oscillations, mV; S_Q is the scale interval of the leak detector (Pa·m^3/s)/mV, calculated by the equation:

$$S_Q=Q_T/\Delta\alpha_T, \tag{12}$$

where Q_T is the value of the test helium leak, Pa·m^3/s; $\Delta\alpha_T$ is the signal output from the test helium leak, mV.
5.2 Record the measured values into protocol: Q_T, Pa·m^3/s; α_f, mV; α_T, mV; $\Delta\alpha_f$, mV; S_q, (Pa·m^3/s)/mV; Q_{min}, Pa·m^3/s.
6 The leak detector pipeline evacuation.
6.1 Open the valve V7 "Input protection".
6.2 Pump down the vacuum chamber to a residual pressure 7 - 8 Pa.
7 The evacuation of the vacuum chamber. The operation is performed for degassing of the measurement object to remove dissolved helium.
7.1 Open the valve V8 for pumping the vacuum chamber.
7.2 Start the stopwatch.
7.3 Make the evacuation by vacuum system at least for 5 hours to achieve the pressure not more than 700-1400 Pa.
7.4 Close valve V8 to stop pumping vacuum chamber.

8 Measuring the helium flow.
8.1 Turn on the measure mode of the leak detector in accordance with its instruction manual [4].
8.2 Open the valve V9.
8.3 Create a pressure differential of 1 atm. from the inner and outer sides of the object, using helium reducer.
8.4 Register the start of the measurements by the stopwatch.
8.5 Record the start time of the leak detector output signal growth.
8.6 Record the signal α_{he} and time in the protocol with measurement interval no more than 5 minutes. The measurement time is not less than 120 minutes.
9 Completion of measurements.
9.1 Close the helium reducer at the end of the measurement time.
9.2 Pump out the helium from vacuum chamber by opening valve V8.
9.3 Turn off the leak detector according to its instruction manual [4].
9.4 Remove the vacuum chamber.
9.5 Remove the presser cover.
9.6 Take away the sample.
10 Determination of flux of the helium through the material, depending on the time.
The determination of the helium flux through the sample is carried out in accordance PNAEG-7-019-89 [2].
10.1 Calculate the value of the helium flux through the sample Q (Pa·m^3/s) by formula:

$$Q = S_Q \cdot \alpha_{he}. \tag{13}$$

Record Q, Pa·m^3/s results to the measurement protocol.
10.2 Calculate the specific flux helium through the sample material Q_{sp}, [Pa·m^3/s]·[m/(m^2·atm)].

$$Q_{sp} = \frac{Q \cdot d}{A \cdot \Delta P_{he}}, \tag{14}$$

where h is the thickness of the sample, A ≈ 1·10^{-3} m^2 – surface area of the sample. Record Q_{sp} results to the measurement protocol.
11 Processing of the measurement results.
Draw the curves of the time dependence:
– the helium flux through the sample Q (Pa·m^3/s),
– the specific flux helium through the sample Q_{sp}, [Pa·m^3/s]·[m/(m^2·atm)].

5. Verification of the measurement of time-dependent permeation rate

The test sample of a polymer material with a known value for the permeability for helium was investigated on the measuring instrument. PTFE was selected as a material for the measurement of permeability for helium.

Permeation coefficient K for helium and PTFE combination is 7.5·10^{-11} m^2/s [8]. The steady state permeation flow Q of a gas exposed with a partial pressure difference 10^5 Pa (1 atm.) to a membrane of thickness d = *1.5* mm and surface area $A \approx 1 \cdot 10^{-3}$ m^2, in accordance with the equation (10), is

$$Q = K\frac{A(p_2 - p_1)}{d} = 7.5 \cdot 10^{-11}\,\frac{\mathrm{m}^2}{\mathrm{s}}\,\frac{1\cdot 10^{-3}\,\mathrm{m}^2 \cdot 10^5\,\mathrm{Pa}}{1.5\cdot 10^{-3}\,\mathrm{m}} = 5 \cdot 10^{-6}\,\frac{\mathrm{Pa}\cdot\mathrm{m}^3}{\mathrm{s}} \tag{15}$$

The comparison of the calculated data with the experimental values, obtained on the device, confirms the accuracy of the measurement apparatus and method. The results are shown in figure 4.

6. Measuring the flux of helium through the polymer materials with metal coatings

Two polymer samples with metal coatings were selected for the measurement:

- Polyethylene. The high molecular weight modified plastelastomer with copper coating 100 nm.
- Low pressure polyethylene HDPE 209-07 with copper coating 100 nm.

The previously evacuated vacuum chamber had been filled to 10^5 Pa with helium, and the flow through the polymer materials was subsequently measured by the leak detector system. The curves of the time dependence of the specific flux helium through the sample Q_{sp} are shown at the figure 5.

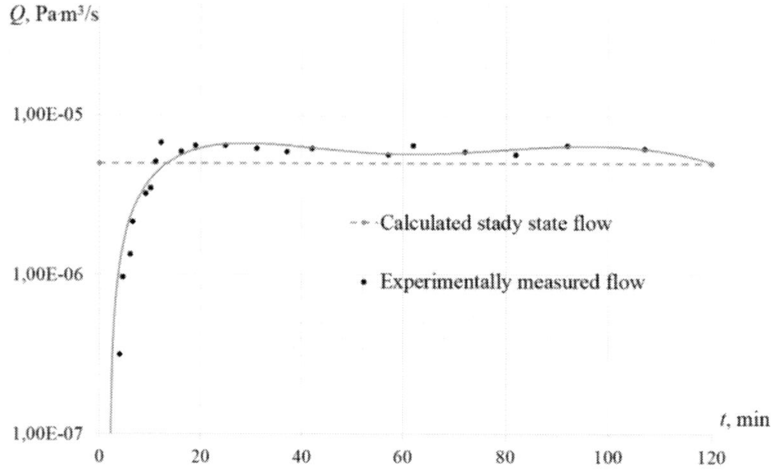

Figure 4. Helium permeation through PTFE sample. Verification of the measurement.

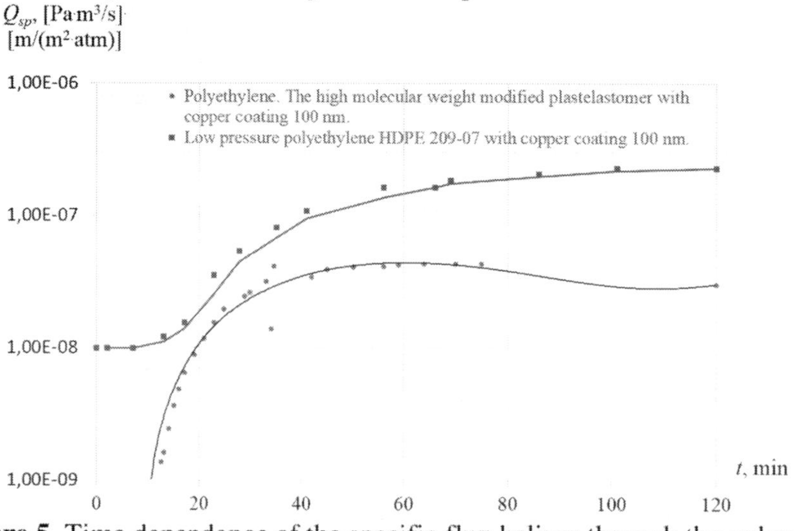

Figure 5. Time dependence of the specific flux helium through the polymers.

It is better to use polyethylene (the high molecular weight modified plastelastomer with copper coating 100 nm.) to create the hermetic devices, in accordance with results of comparison.

7. Conclusions

Apparatus and method for measuring flow of helium through the materials and coatings, obtained by ion-plasma technologies, are developed and tested. The apparatus for the measurement is designed on the basis of a helium leak detector TI1-14, produced by JSC "Zavod Izmeritel", that provides a minimum detectable flow of helium $7 \cdot 10^{-13}$ $Pa \cdot m^3/s$. Research work in the field of diffusion of helium through modern polymer materials with metal coatings are held at this measurement system.

References

[1] OST 5.0170-81. *Non-destructive testing. Metal structures. Gas and liquid leak detection methods.*

[2] PNAEG-7-019-89. *Standardized methodology of control of basic materials (semi-finished products), welded joints and welding equipment and pipelines. Leak detection. Gas and liquid methods.*

[3] GOST 28517-90. *Non-destructive testing. Mass spectrometric method for leak detection. General requirements.*

[4] *Mass spectrometer leak detector TI1-14.* Instructions manual, "Zavod Izmeritel", St. Petersburg, 1991.

[5] OST 11-0808-92 *Non-destructive testing. Leak detection methods.*

[6] Perkins, W. G., *Permeation and outgassing of vacuum materials*, J. Vac. Sci. Technol., 10(4), 543 – 556, 1973.

[7] Barrer, R. M., *Diffusion in and Through Solids*, Cambridge Univ. Press, New York, 1941.

[8] O'Hanlon, J. F., *A User's Guide to Vacuum Technology*, 2nd ed., John Wiley, Hoboken, N. J., 1989.

Hydrogen transport through oxide metal surface under atom and ion irradiation

L Begrambekov[1], O Dvoychenkova[1], A Evsin[1], A Kaplevsky[1], Ya Sadovskiy[1], N. Schitov[2], S Vergasov[1] and D. Yurkov[2]
[1] – National Research Nuclear University MEPhI, Moscow, 115409 Kashirskoe shosse 31, Russian Federation
[2] – Federal State Unitary Enterprise All-Russia Research Institute of Automatics – FSUE VNIIA, Moscow 127055, Sushchevskaya 22, Russian Federation

E-mail: lbb@plasma.mephi.ru

Abstract. Both the latest and earlier achieved results on gas exchange processes on metal surfaces (including stainless steel, titanium, zirconium, tungsten with deposited aluminum oxide coating) under hydrogen atom or plasma irradiation with occasional oxygen impurity are presented in the paper. Mechanisms and regularities of these processes are discussed.
It is demonstrated that surface oxide layer properties as a diffusion barrier strongly depend on external influence on the surface. In particular, it is revealed that low energy hydrogen ion irradiation could slow down hydrogen desorption from metals. Hydrogen atom or ion irradiation combined with simultaneous oxygen admixture accelerates hydrogen desorption from metals.

1. Introduction

Hydrogen content in metals and coatings is usually an unwanted feature that decreases their operational parameters. The negative effects can be worse if an oxide layer is present on the surface that acts as a diffusion barrier for hydrogen atoms. As it has been shown [1], surface oxide layers alter their properties under hydrogen atom or ion irradiation especially when this irradiation is performed with oxygen addition.
The latest and earlier achieved results on gas exchange processes on metal surfaces (including stainless steel, titanium, zirconium, tungsten with deposited aluminum oxide coating) under hydrogen atom or plasma irradiation with occasional oxygen impurity are presented. Mechanisms and regularities of these processes are discussed.

2. Gas exchange processes between hydrogen plasma and stainless steel surface.

Multifunctinoal research apparatus for mass-spectrometry analysis (MIKMA) [1] was used for these and all other experiments presented in the paper. This device allows performing of sample irradiation with hydrogen atoms or plasma ions and gas composition analysis as well as thermal desorption analysis (TDS) of the samples. TDS ramp rate was 5 K/s in all experiments. During irradiation vacuum chamber wall were air-cooled so that their temperature did not exceed 40 C. Residual gas (H_2O ~ 95 at. %, H_2~ 5 at. %) pressure was less than $1x10^{-4}$ Pa. Working gas pressure was 6.6×10^{-1} Pa. At the first stage of experiments the vacuum chamber wall was irradiated with deuterium atoms from D_2+O_2 working gas that were generated by D_2 dissociation at a heated tungsten wire. Then, plasma discharge was initiated in the same gas mixture thus irradiating chamber walls with deuterium ions as well as atoms. At the second stage of experiments deuterium was substituted for hydrogen and vacuum chamber wall was irradiated with hydrogen atoms and ion from H_2+O_2 gas mixture. These series of experiments was performed with various oxygen concentrations: 2, 5, 10, 20 and 30 at%.

Working gas composition change for D_2+10%O_2 gas mixture after cathode heating and plasma initiation are presented at figure 1a. Hydrogen atom irradiation of the chamber walls led to rapid oxygen content decrease in the working gas. It fell below mass spectrometer sensitivity level. At the same time molecules of HD, HDO D_2O appeared in the working gas with simultaneous H_2 and H_2O content increase. When plasma discharge was initiated no further radical changes in gas composition were observed. These results show that hydrogen atom and ion irradiation initiated formation of hydrogen containing molecules on the stainless steel surface. These molecules contained hydrogen from working gas as well as hydrogen from the sorption layer that had diffused from the bulk of the wall.

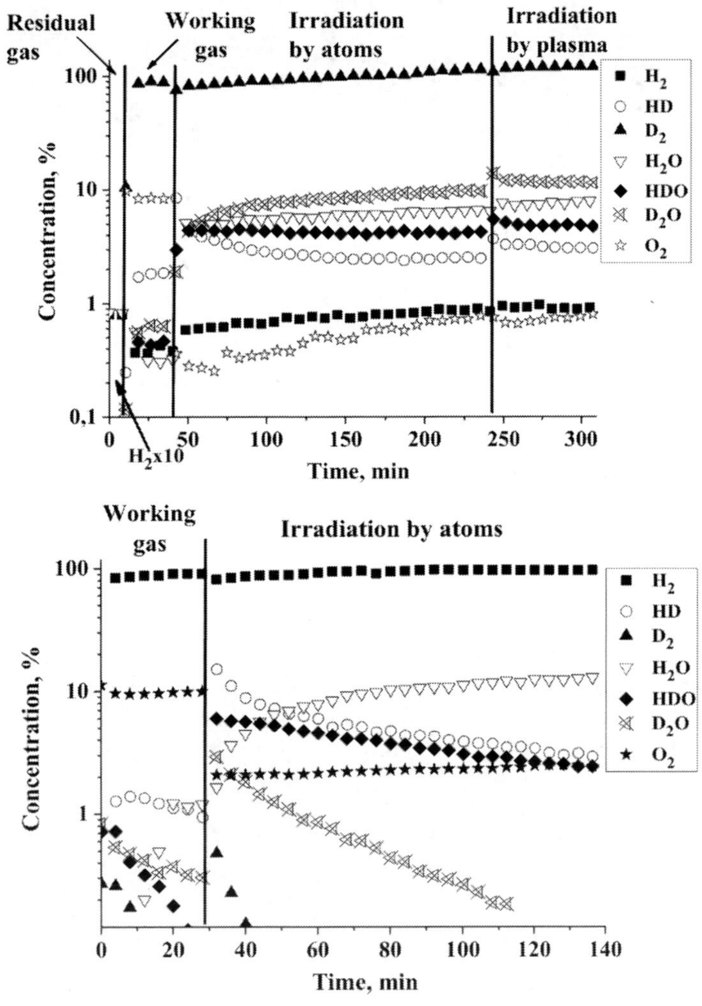

a b

Figure 1. Working gas composition change during: a) D_2+10% O_2 gas mixture, vacuum chamber wall irradiated with atoms and subsequent plasma irradiation, b) H_2+10% O_2 gas mixture with atom irradiation after the wall was irradiated in deuterium plasma.

The processes described above can be supported by a series of exothermic reactions that took place on the surface (Figure 2). Hydrogen atoms and ions initiated these reactions by interacting with chromium oxide (Cr_2O_3) and also participated in subsequent reactions. Oxygen atoms reduced chromium oxide and took part in all "water" molecules generation. At the second stage of experiments hydrogen atom irradiation stimulated deuterium release from the wall which had been trapped during the first stage (Figure 1b).

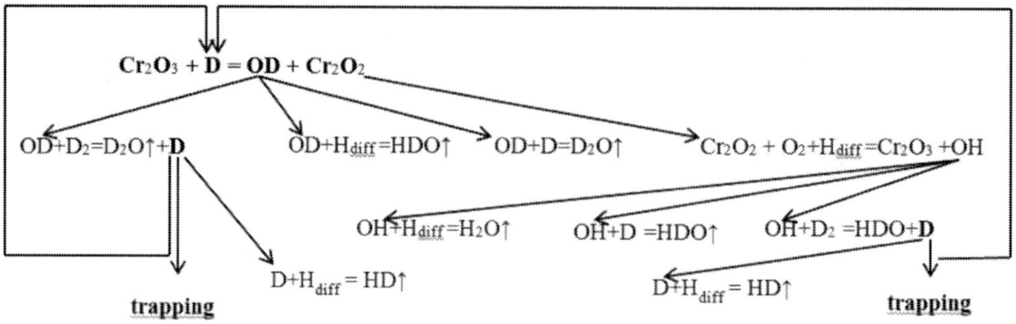

Figure 2. A series of reactions on stainless stell surface under atom irradiation in $D_2 + O_2$ gas mixture that leads to gas composition change and deuterium trapping into the wall

3. Hydrogen transport through the surface under ion irradiation.

Ion irradiation influence on hydrogen transport through 1 μm titanium layer saturated with deuterium at Ti:D=1,7 concentration deposited on molybdenum substrate was studied. The samples were heated to 350 C both with and without hydrogen plasma with oxygen content (0, 2, 4, 10% oxygen) irradiation for the same time. Ion energies were 650 and 3500 eV, ion flux 7×10^{15} at/cm^{2}s.

Deuterium concentrations in the titanium layer before and after the experiments are presented at table 1. It is shown that without any irradiation about half of deuterium is released from the target when it is heated. Deuterium is also released under 3500 eV ion irradiation but no release was registered under 650 eV ion irradiation. In both these cases hydrogen was trapped in the sample. The conclusion can be made that a kind of barrier layer is formed when hydrogen atoms are stopped in the near surface layer.

When oxygen is added to the working gas at first hydrogen trapping increases but after the oxygen concentration reaches 4% it decreases. Deuterium desorption from titanium increased with oxygen content and almost half of deuterium was released when oxygen content was 10%. This shows that after certain amount of oxygen in plasma, surface irradiation stimulates not only deuterium but also hydrogen desorption. Thus, similar processes to those on stainless steel surface are likely to take place.

Table 1. Hydrogen and deuterium concentration change after sample heating at 350 C and after irradiation in (H_2+n%O_2) plasma with 650 or 3500 eV ions at 350 C.

Action	Hydrogen concentration (at/cm^3)	H_{after} / H_{before}	Deuterium concentration (at/cm^3)	D_{after} / D_{before}
Sample before heating	0.6×10^{22}		9.6×10^{22}	
Sample heating without irradiation	0.25×10^{22}	0.4	4.2×10^{22}	0.44
650 eV ion irradiation in H_2 plasma	1.1×10^{22}	1.8	9.6×10^{22}	1
3500 eV ion irradiation in H_2 plasma	0.9×10^{22}	1,5	$3{,}2\times10^{22}$	0,35
650 eV ion irradiation in H_2+2%O_2 plasma	1.2×10^{22}	2.0	9.5×10^{22}	1
650 eV ion irradiation in H_2+4%O_2 plasma	$0{,}9\times10^{22}$	1.5	$8{,}5\times10^{22}$	0.88
3500 eV ion irradiation in H_2+10%O_2 plasma	$0{,}3\times10^{22}$	0.5	$4{,}3\times10^{22}$	0.45

Journal of Physics: Conference Series **567** (2014) 012003 doi:10.1088/1742-6596/567/1/012003

4. Hydrogen transport through zirconium surface under plasma irradiation.

Hydrogen transport in "Э-110" (Zr-1%Nb) zirconium alloy under argon ion irradiation was studied in relation to residual gas pressure. Samples of 7×7×1 mm size were irradiated by 250 eV argon ions with $7{,}5\times10^{15}$ at/cm^{2}s flux. Total fluence was 9×10^{18} at/cm^2, sample temperature 550 K, residual gas pressure was varied between 0,13-4,0 $\times10^{-2}$ Pa (H_2O ~ 92%, H_2 ~ 5%).

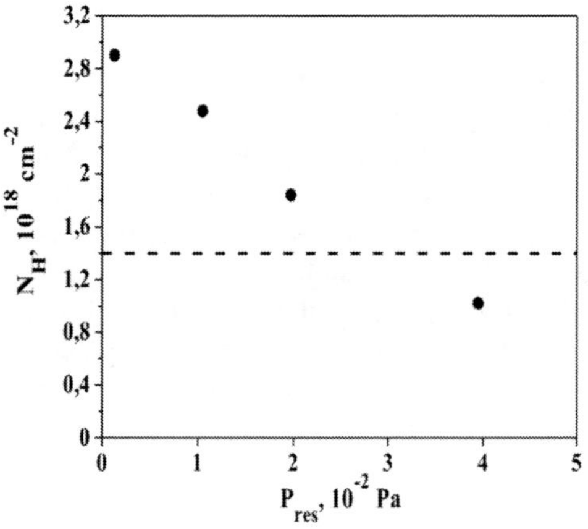

Figure 3. Hydrogen content in irradiated sample versus residual gas pressure. Dotted line marks hydrogen content in non-irradiated sample.

It can be seen (Figure 3) that during irradiation at (1,3–30)x10^{-3} Pa residual pressure hydrogen trapping decreased contrary to irradiation at 3-4 $\times10^{-2}$ Pa residual pressure at which hydrogen desorped from the sample. Preliminary experiments showed that hydrogen desorption from the surface oxide layer occurs at roughly 400 and 1000 K, whereas from the bulk of zirconium at 1300 K. This information allowed to judge on hydrogen trapping sites location based on TDS spectra shape before and after irradiation in argon plasma (Figure 4).

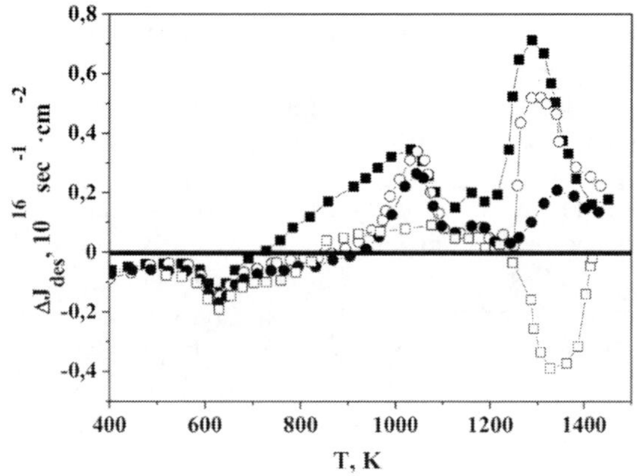

Figure 4. Zirconium alloy (Э-110) TDS spectra difference before and after irradiation in argon plasma at residual pressure of: $1{,}3\cdot10^{-3}$ Pa (■), $1{,}1\cdot10^{-2}$ Pa (○), $2\cdot10^{-2}$ Pa (●), $4\cdot10^{-2}$ Pa (□)

At (1.3-30) $\times 10^{-3}$ Pa residual pressure hydrogen is seen to be trapped both in the surface oxide layer and in the bulk of the material. At higher residual gas pressures less hydrogen is trapped in the bulk. One may suppose that oxygen atoms generated from water molecules in plasma stimulated similar to described above processes on the surface of zirconium.

5. Barrier properties of aluminum oxide layer on tungsten surface.

Behavior of hydrogen and deuterium atoms trapped in tungsten after depositing aluminum oxide coating layer was studied. 300 eV deuterium ions were irradiating the sample with $7{,}3\times 10^{15}$ at/cm^{2}s flux, $2{,}6\times 10^{19}$ at/cm^2 total fluence at 500 K sample temperature. After that, aluminum oxide coating was formed on both sides of the sample by sputtering an aluminum target in Ar+n%O_2 gas mixture (n=1-8). During deposition sample temperature was 450-500 K, resulting coating thickness was 200-700 nm, deposition duration 1 h per each side of the sample.
It is shown (Table 2), that during deposition in plasma with less 1% oxygen addition deuterium content in the sample decreased by roughly 50%. In 4 and 8% oxygen addition cases ~90% of deuterium was released. The behavior of "technological" hydrogen present in the sample was practically the same.

Table 2. Hydrogen and deuterium desorption from tungsten samples after aluminum oxide coating deposition.

Oxygen concentration in plasma	Hydrogen desorption (at/cm^2)	Deuterium desorption (at/cm^2)	$(H_{before}-H_{after})/H_{before}$	$(D_{before}-D_{after})/D_{before}$
<1%	2.2×10^{17}	1.5×10^{16}	≈13%	≈21%
4%	5.4×10^{16}	3.2×10^{15}	≈68%	≈82%
8%	7.1×10^{16}	3.5×10^{15}	≈75%	≈83%

Similarity of the results obtained in all these experiments could be connected with the similarity in experimental conditions. In all cases an oxide layer on the surface was irradiated by ion from plasma containing both hydrogen and oxygen atoms. Only the surface of the sample was affected, with no direct influence on hydrogen diffusion through the bulk of material or oxide layer.

6. Summary

Surface oxide layer properties as a diffusion barrier on metals strongly depend on external influence. Slow hydrogen atoms or ions that are stopped in the oxide layer or just beyond it slow down hydrogen diffusion to the surface. High energy ions that are stopped further in the bulk increase hydrogen concentration in irradiated material. Thus hydrogen diffusion to the surface and its desorption is accelerated.
When oxygen is added to the working gas slow hydrogen atoms and ions initiate a series of exothermic reactions involving surface oxide layer and hydrogen from the bulk of material. As a result of these reactions hydrogen atoms are associated in hydrogen-containing molecules that are released from the surface. Acceleration of hydrogen diffusion from the bulk of material as a result of these processes shows that hydrogen atom relaxation on the surface is the main factor defining barrier properties of surface oxide layers. Acceleration of relaxation activates hydrogen removal from metals.

Acknowledgements
This work was supported by Competitiveness Growth Program of the Federal Autonomous Educational Institution of Higher Professional Education National Research Nuclear University MEPhI.

References

[1] A. Airapetov, L. Begrambekov, S. Bremond, et al. J. of Nucl. Mater. 415(2011)1042-1045

The model of the positive column of a glow discharge with the influence of the acoustic oscillations

N F Kashapov, A I Saifutdinov and S A Fadeev
Kazan Federal University, 18 Kremlyovskaya str., Kazan 420008, Russian Federation

E-mail: Sergey_Fadeev@kpfu.ru

Abstract. In article the model of the positive column of a glow discharge with the influence of acoustic waves is developed. It is shown that the radial convective flow caused by the acoustic streaming leads to a redistribution of the gas temperature, and reduces its gradient between the axis and the wall of the discharge chamber.

1. Introduction

The low temperature plasma of glow discharge is a partially ionized gas with a large electron temperature and has a wide range of practical applications. The plasma of this type used in MHD generators [1] and magnetron sputtering system [2], in the processes of plasma modification of surface properties [3-6] and in the plasma antennas [7]. We should also mention the gas CO_2 lasers [1, 8], which are the most widely used lasers in material processing. The most powerful lasers have a wide range of technological applications. The most reasonable way to improve the efficiency of CO_2 lasers is to increase the pressure in the discharge chamber and the power into the active medium of the laser. However, in the process of improving these parameters glow discharge of the CO_2 laser becomes unstable. He loses uniformity and constricted. This leads to a complete cessation of radiation generation. Overcoming this kind of instability is the main problem in creating high-power lasers.

In each of the kinds of practical application of low temperature plasma can identify processes which are the result of the interaction of plasma and external factors. Most detailed investigated the influence of electromagnetic waves, electric fields, magnetic fields, electron beams, ion beams, neutral beams, laminar and turbulent flows on the plasma properties. During the last decade intensively investigated issues of interaction of acoustic waves with plasma. Influence of acoustic waves on the low-temperature plasma is accompanied by a variety of processes [9, 10, 11]: lowering the temperature of the gas, reduction of the radial gradient of the gas temperature, the increase of the longitudinal electric field, raising the temperature of the electrons, modulation of the discharge current and voltage. It can be effectively used in the development of high-power CO_2 lasers. For a description of these phenomena is necessary to develop a mathematical model of the positive column of a glow discharge taking into account influence of acoustic waves.

2. The model of the positive column of a glow discharge with the influence of acoustic streaming

We know that the acoustic low-frequency oscillations cannot influence the transitions between electronic levels. The phenomena occurring in the discharge chamber at excitation of acoustic waves associated with the hydrodynamic flows, which affect the structure and stability of the positive column of glow discharge. This leads to a change the processes of ionization and recombination.

In this case, acoustic oscillations in a glow discharge plasma can be accounted through the vortex the so-called acoustic streaming. Acoustic streaming excited, firstly, due to the boundary layer at the wall where the velocity is reduced from the value in the acoustic waves in the middle of the tube to zero at the wall of the tube, and secondly, because of the strong temperature gradient along the radius of the tube.

As shown in [9], acoustic flows generated in the field of the acoustic wave and field of velocities can be expressed as follows (in the cylindrical coordinate system)

$$u_r = -\frac{u_a^2}{10c_0}\frac{R^2}{\delta^2}\frac{r}{\lambda}\left(1-\frac{r^2}{R^2}\right)\cos 2kx, \tag{1}$$

$$u_x = -\frac{u_a^2}{c_0}\frac{\alpha}{120}\frac{R^2}{\delta^2}\frac{r}{\lambda}\left(1-\frac{r^2}{R^2}\right)\left(1-\frac{5r^2}{R^2}\right)\sin 2kx, \tag{2}$$

where u_a is the parameter characterizing the temperature inhomogeneity of the medium

$$u_a = -\frac{P_A}{\rho c_0}\left(1+\frac{\alpha\omega^2}{c_0^2}r^2\left(1+\frac{r^2}{2R^2}\right)\right). \tag{3}$$

Here R is the tube's radius, α is the size of the inhomogeneity in the acoustic field, ρ is the gas density, ω is the frequency of the acoustic wave, c_0 is the speed at which the acoustic wave travels in an undistorted medium, δ is the size of the boundary layer, λ is wavelength, P_A is pressure amplitude of the acoustic oscillations in the tube, k is wave vector.

A simplified model of the spatial structure of the positive column of a glow discharge with cylindrical symmetry (from the axis to the wall of the discharge chamber) with the influence of acoustic flows leading to decontraction discharge can be formulated on the basis of the following system of equations

1) Balance equation for the concentration of charged particles (for electrically neutral plasma)

$$\frac{u_x}{R}\frac{\partial n}{\partial x}+\frac{u_r}{R}\frac{\partial n}{\partial r}=\frac{D_a}{R^2 r}\frac{\partial}{\partial r}\left(r\frac{\partial n}{\partial r}\right)+\nu_i n-\beta n^2, \tag{4}$$

where ν_i – is the ionization frequency, which is a function of the electric field E, β is the recombination coefficien, D_a is the ambipolar diffusion coefficient. The boundary conditions for equation (4) are given on the chamber wall and the discharge axis

$$n(R) = n'(0) = 0. \tag{5}$$

2) Energy balance equation for neutral particles

$$\rho c_p\frac{u_x}{R}\frac{\partial T}{\partial x}+\rho c_p\frac{u_r}{R}\frac{\partial T}{\partial r}=\frac{\chi}{R^2 r}\frac{\partial}{\partial r}\left(r\frac{\partial T}{\partial r}\right)+jE, \tag{6}$$

where χ is coefficient of thermal conductivity, c_p is thermal capacity, j is current density. The equation describes the balance of Joule heating of gas and heat loss to the walls by thermal flow with a coefficient of thermal conductivity χ and by acoustic streaming. Boundary conditions for (6) are given

$$T(R) = T_0, T'(0) = 0 \tag{7}$$

3) The equation of state of the gas in which the pressure is determined by the temperature of the neutral particles

$$p = N\kappa T, \tag{8}$$

where N is concentration of neutral particles, κ is Boltzmann constant. The equation describes the redistribution of neutral particles N over the radius caused by nonuniform heating of the gas.

4) The equation for the discharge current (current is determined by the electrons)

$$i = e\mu_e E \int_0^R n(r) r dr, \tag{9}$$

which binds the concentration of charged particles with external parameters. Here is the mobility coefficient.

3. Evaluation of the radial distribution parameters of a glow discharge with acoustic flows

To solve system of equations (4)-(9) with (1) and (2) requires the use of special numerical methods. However, approximate analytical evaluation of the radial distribution of the parameters of the positive column, which uncontracted by creating acoustic flows in some approximation can be determined, assuming that the radial component of velocity u_r is constant, and distribution of the plasma parameters along the x axis uniform. In addition, we assume that, under our pressure due to acoustic streaming, loss of charged particles at the chamber walls dominates the recombination in the plasma volume.

In this case, the system of equations can be written as

$$\frac{u_r}{R}\frac{\partial n}{\partial r} = \frac{D_a}{R^2 r}\frac{\partial}{\partial r}\left(r\frac{\partial n}{\partial r}\right) + \nu_i n, \tag{10}$$

$$\rho c_p \frac{u_r}{R}\frac{\partial T}{\partial r} = \frac{\chi}{R^2 r}\frac{\partial}{\partial r}\left(r\frac{\partial T}{\partial r}\right) + jE, \tag{11}$$

$$p = N\kappa T \tag{12}$$

$$i = e\mu_e E \int_0^R n(r) r dr \tag{13}$$

Equation (9) is the confluent hypergeometric equation and its solution has the form

$$n(r) = n_0 \exp\left(-\frac{r\left(\sqrt{u_r^2 - 4\nu_i D_a} - u_r\right)}{2D_a}\right)\Phi(a, b, \xi), \tag{14}$$

where n_0 is electron density on the axis of symmetry,

$$\Phi(a, b, \xi) = \frac{\Gamma(b)}{\Gamma(a)\Gamma(a-b)}\int_0^1 e^{\xi t} t^{a-1}(1-t)^{b-a-1} dt,$$

is special function of Kummer [12],

$$\Gamma(r) = \int_0^\infty e^{-t} t^{r-1} dt,$$

is gamma function,

$$a = \frac{\left(\sqrt{u_r^2 - 4v_i D_a} - u_r\right)}{2\sqrt{u_r^2 - 4v_i D_a}},\ b = 1,\ \xi = \frac{r\sqrt{u_r^2 - 4v_i D_a}}{D_a}.$$

The solution of equation (11) can be found by direct integration

$$T(r) = -\int_0^r \left(\frac{R^2}{\chi} \exp\left(\frac{\rho c_p u_r R r}{\chi}\right) \int_0^r j(r)E(r) \exp\left(\frac{\rho c_p u_r r}{\chi}\right) r dr\right) dr + T_0 \tag{15}$$

Profile of the radial distribution of the electric field uncontracted column can be found by substituting the expression (14) in (13) and solving an integral equation relatively E, or by equating the sum diffusion and convection components with drift component of the radial flow of electrons (Boltzmann equilibrium)

$$-D_e \frac{dn_e}{dr} + u_r \frac{dn_e}{dr} - \mu_e E_r n_e = 0, \tag{16}$$

then

$$E_r = -\frac{(D_e - u_r)}{\mu_e} \frac{d}{dr} \ln\left(n_0 \exp\left(-\frac{r\left(\sqrt{u_r^2 - 4v_i D_a} - u_r\right)}{2D_a} \right) \Phi(a, b, \xi) \right). \tag{17}$$

4. Conclusions

In a glow discharge plasma parameters are changed due to the presence of acoustic oscillations. To describe such processes developed a mathematical model of the positive column of a glow discharge with the influence of acoustic streaming. Revealed (equation 15) that the convective radial flow caused by acoustic streaming causes a redistribution of the gas temperature, to reduce its gradient between the axis and the wall of the discharge chamber, which entails a redistribution of the concentration of neutral particles (equation 12). The uniform distribution of the density of neutral particles, as well as convective concentration of charged particles to the walls of the chamber should lead to a discontraction of positive column.

The work is performed according to the Russian Government Program of Competitive Growth of Kazan Federal University.

References

[1] Raizer Y P 1992 *Gas Discharge Physics* (Moscow: "Science") p 536
[2] Galyautdinov R T and Kashapov N F 2003 *Svarochnoe Proizvodstvo* **(3)** 27–31
[3] Abdullin I Sh, Galyautdinov R T and Kashapov N F 2001 *Inzhenerno-Fizicheskii Zhurnal* **74** 104–107
[4] Galyautdinov R T, Kasparov N F and Luchkin G S 2002 *Inzhenerno-Fizicheskii Zhurnal* **75** 170–173
[5] Gavrilova V A, Fazlyyyakhmatov M G and Kashapov N F 2013 *J. Phys.: Conf. Series* **479** 012010
[6] Dautov G, Fayrushin I 2013 *J. Phys.: Conf. Series* **479** 012013
[7] Gusein-Zade N G, Minaev I M, Rukhadze A A and Rukhadze K Z *Journal of Communications Technology and Electronic* **56** 1207–11
[8] Kashapov N F and Israfilov Z K 1991 *Journal of Engineering Physics* **60** 364–368
[9] Aramyan A R and Galechyan G A 2007 *Physics-Uspekhi* **50** 1147–69

[10] Fadeev S A and Kashapov N F 2013 *J. Phys.: Conf. Series* **479** 012009
[11] Israfilov Z K and Kashapov N F 1987 *Soviet Aeronautics* **30** 97–99
[12] Polyanin A D and Zaitsev V F 2003 *Handbook of Exact Solutions for Ordinary Differential Equations* (Boca Raton: A CRC Press Company)

Calculation of distribution of potential near the surface of metal particle in the dust-electron thermal plasma

G Dautov[1], I Fayrushin[2], N Kashapov[2], and I Dautov[2]
[1]General Physics Department, Kazan National Research Technical University by named A.N. Tupolev-KAI, Kazan, 420111, Russia
[2]Department of Thechnical Physics and Power Engineering, Kazan Federal University, Kazan, 420008, Russia

E-mail: fairushin_ilnaz@mail.ru

Abstract. We obtained the equation, which describes the distribution of the potential in an equilibrium dust-electron plasma taking into account parameters of the electron gas inside the dust particles. The inclusion of these parameters performed on the basis of the model of "solid-state plasma," considering the condensed particle system as the ion core and the free electron gas. The analytical expressions for the potential distributions in thermal equilibrium dust-electron plasma were obtained, using a number of simplifying assumptions. It is found that near the particle surface there is a large gradient of the electric potential and as a result the concentration of free electrons, i.e. electrons in the dust particle are located in a potential well. It is shown that the form of the potential barrier at the particle surface depends on the plasma temperature, particle size and concentration.

1. Introduction

One of the most important characteristics of dust-electron thermal plasma are potential and electron density distributions [1-7]. To determine these characteristics of the data which are used in this paper the model of "jelly", which previously has been widely used in the description of the electrical properties of atomic clusters [8-10]. According to this model, the dust particle is represented as two-components system. The first component is ion core, which create uniform positive background in the whole volume of the dust particle. The second component is the electronic gas. The density of this gas distribute in the area from the condition of equilibrium of forces internal pressure of the gas and electrostatic forces. Thus, in this model, a dust particle can be regarded as the area of solid state plasma [11-13]. From the above it follows that the distribution of the potential and the electron density within and around the dust particles are dependent on the temperature, type of the substance, size and the concentration of particles.

2. Theoretical background and results

Consider the system of equations describing the potential distribution and the electron density in solid-state plasma and in the space around it in a state of statistical equilibrium when the temperature is low and can be neglected of the ionization of the gas.

It is known from the statistical physics, the properties of the electron gas are described by the Fermi-Dirac distribution

Journal of Physics: Conference Series **567** (2014) 012005 doi:10.1088/1742-6596/567/1/012005

$$n_j = \frac{1}{1+e^{\frac{\varepsilon_j-\mu}{\theta}}}, \tag{1}$$

where θ – statistical temperature equal kT, k - Boltzmann constant, T- absolute temperature, μ – Fermi energy, ε_j - electron energy in the j -th quantum state, n_j - average numbers of electrons in the same quantum state with energy of ε_j.

In the semiclassical approximation, the number of electrons per unit volume in the energy range from ε to ε + dε defined by the formula [14]

$$n_e(\varepsilon)=\int_0^{\infty}\frac{4\pi\cdot(2m_e)^{\frac{3}{2}}\varepsilon^{\frac{1}{2}}}{h^3\left(1+e^{\frac{\varepsilon-\mu}{\theta}}\right)}\cdot d\varepsilon. \tag{2}$$

Here h – Planck's constant, m_e - mass of the electron, n_e - electron density.

When θ = 0, the electron gas is completely degenerate and the equation (2) is obtained from the Fermi energy:

$$\mu=\frac{h^2}{2m_e}\cdot\left(\frac{3n_e}{8\pi}\right)^{\frac{2}{3}}. \tag{3}$$

The distribution of the potential φ of the electrostatic field is described by Poisson equation

$$\Delta\varphi=\frac{n_e-n_i}{\varepsilon_0\varepsilon_1}q, \tag{5}$$

where ε_1 - relative permittivity, ε_0 - electric constant, n_i- the concentration of positive charge of the ion core (the density of the positive background), q - the absolute value of the electron charge.

The condition of statistical equilibrium of the electron gas can be written as[14]

$$\mu-q\varphi=const. \tag{6}$$

Equations (2), (5) and (6) give the boundary conditions completely determine the spatial distribution of the potential and the electron density.

The system of equations (2), (5) and (6) can be obtained by a single equation, which calculates the potential distribution and electron density. From (6) wefind

$$\Delta\varphi=\frac{\Delta\mu}{q}. \tag{7}$$

Substitution of expressions (2) and (7) into (5) leads to a nonlinear integro-differential equation

$$\Delta\mu = \frac{q^2}{\varepsilon_0\varepsilon_1}\left[\int_0^\infty \frac{4\pi(2m_e)^{\frac{3}{2}}\varepsilon^{\frac{1}{2}}}{h^3\left(1+e^{\frac{\varepsilon-\mu}{\theta}}\right)}d\varepsilon - n_i\right]. \quad (8)$$

This equation is applicable to both degenerate and non-degenerate electron gas. This is particularly important in the case of dust particles from the metal. In this case, there are areas of the degenerate and non-degenerate electron gas.

In the case of degenerationgas (2) we find

$$n_e = \frac{16\sqrt{2}\pi m^{3/2}}{3h^3}\mu^{3/2}. \quad (9)$$

Substituting this expression in (8), we obtain

$$\Delta\mu = \frac{q^2}{\varepsilon_0\varepsilon_1}\left(\frac{16\sqrt{2}\pi m^{3/2}}{3h^3}\mu^{3/2} - n_i\right). \quad (10)$$

If we introduce the dimensionless potential $\psi = \frac{q\varphi}{\mu_0}$, for the case of spherical symmetry from the (12) and (13) we obtain the equation

$$\frac{1}{x^2}\cdot\frac{d}{dx}\left(x^2\cdot\frac{d\psi}{dx}\right) = a^2\left[(1+\psi)^{3/2} - \bar{n}_i\right], \quad (11)$$

where $x = \frac{r}{R}$, $\frac{n_i}{n_{e0}} = \bar{n}_i$, $a^2 = \frac{q^2R^2n_{e0}}{\mu_0\varepsilon_1\varepsilon_0}$.

Consider the case, when $|\psi| < 1$. We expand the value $(1+\psi)^{3/2}$ in the seriesand retain the first two terms: $(1+\psi)^{3/2} = 1 + \frac{3}{2}\psi$.

Given of this expression of the equation(15) obtained by

$$\frac{1}{x^2}\cdot\frac{d}{dx}\left(x^2\cdot\frac{d\psi}{dx}\right) = a^2\left(1 + \frac{3}{2}\psi - \bar{n}_i\right). \quad (12)$$

The initial conditions for this equation is $\psi(0)=0$, $\psi'(0)=0$. Its solution is

$$\psi(x)=\frac{2}{3}\cdot\frac{(a-1)\cdot e^{a}-\left(\frac{a\lambda-1}{a\lambda+1}\right)\cdot(a+1)\cdot e^{a(2\lambda-1)}}{2a\cdot\left[(a-1)\cdot e^{a}-(a+1)\cdot\left(\frac{a\lambda-1}{a\lambda+1}\right)\cdot e^{a(2\lambda-1)}+\left(\frac{a\lambda-1}{a\lambda+1}\right)\cdot e^{2a\lambda}+1\right]}\cdot\left[\frac{\left(e^{ax}-e^{-ax}\right)}{x}-2a\right],\text{if } 0\le x\le 1 \quad (13)$$

If the concentration of dust particles is large; concentration of electrons in the region $1<x\leq\lambda$ will be large and the electron gas is degenerate. In this case, the potential distribution in the area $1<x\leq\lambda$ defined by the formula.

$$\psi(x)=\frac{2}{3}\cdot\left(\frac{(a-1)\cdot e^{a}+(a+1)\cdot e^{-a}}{2a\cdot\left((a-1)\cdot e^{a}-(a+1)\cdot\left(\frac{a\lambda-1}{a\lambda+1}\right)\cdot e^{a(2\lambda-1)}+\left(\frac{a\lambda-1}{a\lambda+1}\right)\cdot e^{2a\lambda}+1\right)}\cdot\frac{e^{ax}+\left(\frac{a\lambda-1}{a\lambda+1}\right)\cdot e^{a(2\lambda-x)}}{x}-1\right),\text{ if } 1<x\leq\lambda \quad (14)$$

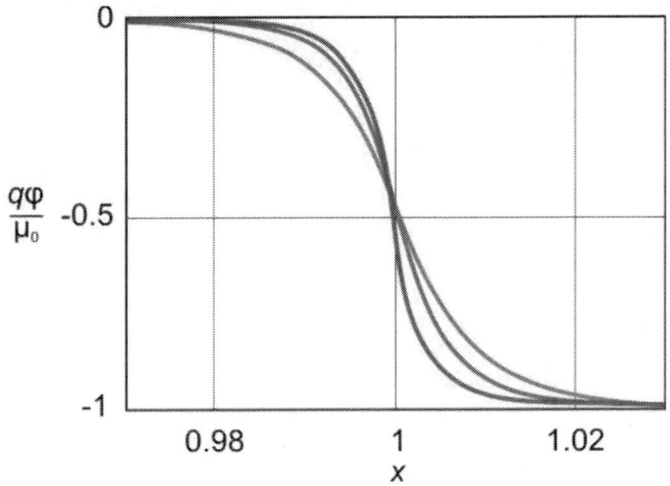

Fig. Distribution the dimesionless potential at $n_i = 10^{27}\,\text{m}^{-3}$ and different of the radius of the dust particle ($R=2\cdot10^{-8}$ m – green line, $R=3\cdot10^{-8}$ m – blue line, $R=4\cdot10^{-8}$ m – red line).

Figure shows plots of the potential distributions at the particle surface, constructed by the formulas (13) and (14). As can be seen from this figure, the potential drop occurs at the surface of the particles at distances comparable to the interatomic. Depending on the particle radius distance data differ. With decreasing particle size increases the distance over which the potential drops, which can be explained by the high specific output of electrons beyond the dust particle, when the radius is decreases.

If the concentration of dust particles is low, in the direction of the axis x degeneracy of the electron gas is gradually removed. In this case, the potential distribution is found by solving equation (8), which is valid for both case degenerate and non-degenerate electron gas.

In conclusion, we obtained the equation (8), which describes the distribution of the potential and the electron density in an equilibrium dust-electron plasma taking into account parameters of the electron gas inside the dust particles. The inclusion of these parameters performed on the basis of the model of "solid-state plasma", considering the condensed particle system as the ion core and the free electron gas. Solving equation (8) using the number of simplifying assumptions, we have been obtained the analytical expressions for the potential distributions and in thermal equilibrium dust-electron plasma. We have been found that near the particle surface there is a large gradient of the electric potential and as a result, the concentration of free electrons, i.e. electrons in the dust particle are located in a potential well. The obtained expressions allow to calculate the influence of several parameters such as temperature, concentration and dust particle size, type of the substance of particles on properties of the equilibrium dust-electron plasma.

Acknowledgements
The work is performed according to the Russian Government Program of Competitive Growth of Kazan Federal University.

References

[1] Complex and Dusty Plasmas, edited by V. E. Fortov and G. E. Morfill (CRC Press, Boca Raton, FL, 2010).
[2] V.I. Vishnyakov, Phy. Rev. E **85**, 026402 (2012).
[3] V.I. Vishnyakov and G.S. Dragan, Phys. Rev. E **73**, 026403 (2006).
[4] V.I. Vishnyakov, G.S. Dragan and V.M. Evtuhov, Phys. Rev. E **76**, 036402 (2007).
[5] Dautov G, Kashapov N, Larionov V, Fayrushin I et al. 2013 J. Phys.: Conf. Series **479** 011001
[6] Dautov G, Fayrushin I 2013 J. Phys.: Conf. Series **479** 012013
[7] Dautov G, Dautov I, Fayrushin I and Kashapov N 2013 J. Phys.: Conf. Series **479** 012014
[8] Dautov G, Dautov I, Fayrushin I and Kashapov N 2013 J. Phys.: Conf. Series **479** 012001
[9] R.D. Smirnov, A.Y. Pigarov, M. Rosenberg, S.I. Krasheninnikov, and D.A. Mendis, Plasma Phys. Controlled Fusion **49**, 347 (2007).
[10] J. Vaverka, I. Richterová, M. Vyšinka, J. Pavlü, J. Šafránková, and Z. Němeček, Plasma Phys. Controlled Fusion 56, 025001 (2014).
[11] V.I. Krauz, Yu.V. Martynenko, N.YuSvechnikov, V.P. Smirnov, V.G. Stankevich, L.N. Khimchenko, Phys. Usp.**53**1015–1038 (2010).
[12] W. Ekardt, Phys. Rev. B. 29, 1558 (1984).
[13] V.K. lvanov, A.N. Ipatov, V.A. Kharchenko,JETP, Vol. 82, No 3, p. 485 (1996)
[14] M.B. Smirnov, V.P. Krainov, JETP, Vol. 88, No 6, p. 1102 (1999)
[15] N.W. Ashcroft and N. D. Mermin, Solid State Physics (Saunders College, Philadelphia, 1976).
[16] L.A. Artsimovich, R.Z. Sagdeev, Plasma physics for physicists. Moscow, Atomizdat, 1979. 320 p.
[17] B.M. Smirnov, Phys. Usp. **45** 1251–1286 (2002).
[18] L. D. Landau and E. M. Lifshitz, *Statisticheskaya Fizika*(Nauka, Moscow, 1976), Vol. 1 [*Statistical Physics* (Pergamon Press, Oxford, 1980)]

Calculation of distribution of potential and electron concentration in the dust-electron thermal plasma with the axial geometry particles

G Dautov[1], I Fayrushin[2] and N Kashapov[2]
[1]General Physics Department, Kazan National Research Technical University by named A.N. Tupolev-KAI, Kazan, 420111, Russia
[2]Department of Thechnical Physics and Power Engineering, Kazan Federal University, Kazan, 420008, Russia

E-mail: fairushin_ilnaz@mail.ru

Abstract. We obtained the equation, which describes the distribution of the potential and the electron density in an equilibrium dust-electron plasma taking into account parameters of the electron gas inside the axial geomtry dust particles. The inclusion of these parameters performed on the basis of the model of "solid-state plasma," considering the condensed particle system as the ion core and the free electron gas.

1. Introduction

The system which consists of dust particles, free electrons and neutral gas atoms at atmospheric or higher pressure at temperatures of 1000-3000 K called dust-electron thermal plasma [1-8]. Under these conditions, this system is an isothermal consequence of the heat of intense collisional, i.e. established the statistical equilibrium. Such a plasma is formed during the combustion of solid and liquid fuels, in the process of thermal spray coating, in the channels of the MHD generator, a plasma chemical reactors, the synthesis of nano- and micro-sized particles and magnetic fusion devices [9-11]. The characteristic size of the dust particles is in the range of 0.01 to 100 microns. Heated dust grains are a source of free electrons due to emission processes. As a consequence, the dust particles are positively charged. If such a plasma does not contain easily ionizable impurities of alkali metal, can be neglected by the presence of the ions.

One of the most important characteristics of dust-electron thermal plasma are potential and electron density distributions. To determine these characteristics of the data which are used in this paper the model of "jelly", which previously has been widely used in the description of the electrical properties of atomic clusters [12-14]. According to this model, the dust particle is represented as two-components system. The first component is ion core, which create uniform positive background in the whole volume of the dust particle. The second component is the electronic gas. The density of this gas distribute in the area from the condition of equilibrium of forces internal pressure of the gas and electrostatic forces. Thus, in this model, a dust particle can be regarded as the area of solid state plasma [15-17]. From the above it follows that the distribution of the potential and the electron density within and around the dust particles are dependent on the temperature, type of the substance, size and the concentration of particles.

2. Theoretical background and results

Consider the system of equations describing the potential distribution and the electron density in solid-state plasma and in the space around it in a state of statistical equilibrium [18] when the temperature is low and can be neglected of the ionization of the gas.

The distribution of the potential φ of the electrostatic field is described by Poisson equation

$$\Delta\varphi = \frac{n_e - n_i}{\varepsilon_0 \varepsilon_1} q, \tag{1}$$

where ε_1 - relative permittivity, ε_0 - electric constant, n_i - the concentration of positive charge of the ion core (the density of the positive background), q - the absolute value of the electron charge.

From the Boltzmann formula

$$n_e = n_{e0} e^{\frac{q\varphi}{\theta}}, \tag{2}$$

where n_{e0} - the electron density at the origin of the potential, i.e. when φ=0, we obtain the Poisson - Boltzmann equation

$$\Delta\varphi = \frac{n_{e0} e^{\frac{q\varphi}{\theta}} - n_i}{\varepsilon_0 \varepsilon_1} q. \tag{3}$$

The solution of equations (2) and (3) determines the distribution of the electron density. Consider a simple task when the dust particles are identical cylinders and these concentration will be constant in the space.

In the case of axial symmetry, of the equation (3) can be written as

$$\frac{\varepsilon_0 \varepsilon_1}{r} \cdot \frac{d}{dr}\left(r \cdot \frac{d\varphi}{dr} \right) = \left(n_{e0} e^{\frac{q\varphi}{\theta}} - n_i \right) q. \tag{4}$$

As a result of the introduction of dimensionless quantities: $\frac{r}{R} = x$, $\frac{q\varphi}{kT} = \phi$, $\frac{n_i}{n_{e0}} = \overline{n_i}$, we obtain

$$\frac{1}{x} \cdot \frac{d}{dx}\left(x \cdot \frac{d\phi}{dx} \right) = b^2 \left(e^{\phi} - \overline{n_i} \right),\ b^2 = \frac{q^2 R^2 n_{e0}}{kT \varepsilon_1 \varepsilon_0}. \tag{5}$$

Here R – is a radius of the dust particle, n_{e0} - the electron density in the center of the cylinder, $n_i = const$ in the field $0 \le x \le 1$ and $n_i = 0$ in the field of $x > 1$. Some of electrons leave from dust particles and therefore $n_i > n_{e0}$. The equation (11) will be solved under the conditions

$$\phi(0) = 0,\ \phi'(0) = 0,\ \phi'(\lambda) = 0,$$

where $\lambda = l/R$, $2l$ - the distance between the centers of two neighboring dust particles. The last boundary condition implies the vanishing of the total electric charge in the volume bounded by a sphere of radius l and is used to determine the value of n_i.

Figure 1 shows plots of the electric potential distribution along the radius for different values of the absolute temperature. As can be seen in the dust particle with increasing x potential first decreases slowly and then approaching the particle surface decreases rapidly. This is the due of the emergence of a large gradient of the electron density near the surface. Accordingly, there is a large electric field. With increasing temperature, the potential drop on the whole interval from 0 to 1 decreases owing to an increase in the electron yield of particles.

Journal of Physics: Conference Series **567** (2014) 012006 doi:10.1088/1742-6596/567/1/012006

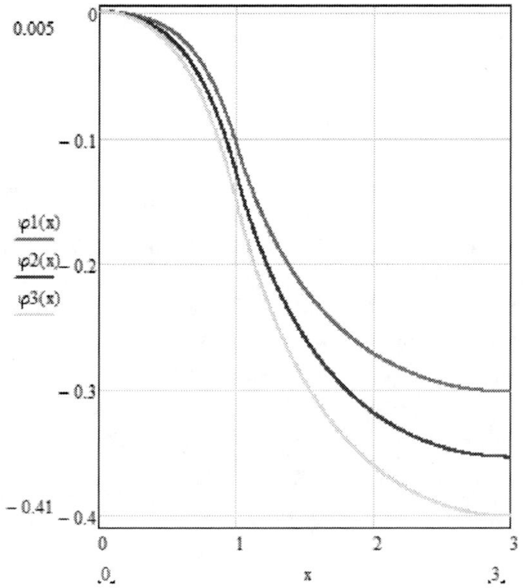

FIG. 1 Distribution of the potential at $R = 10^{-6}$ m, $n_i = 10^{19}$ m^{-3}, and at different temperatures (1000 K - red line, 1250 K - blue line, 1500 K - green line).

Since the potential and the electron density are is directly related to the Boltzmann formula, x increases in $x \approx 1$ magnitude n_e decreases sharply. With the removal of the particles decreases and the rate of change at the point of the condition $\frac{dn_e}{dx} = 0$ (see Fig. 2).

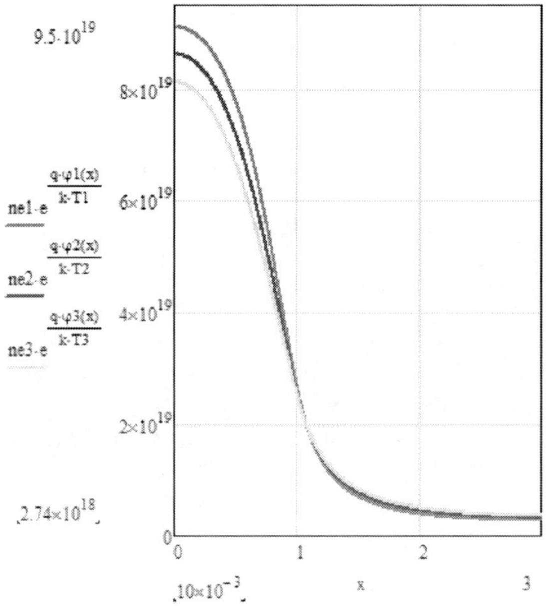

Fig. 2. Distribution of the electron density at $R = 10^{-6}$ m, $n_i = 10^{19}$ m^{-3}, and at different temperatures (1000 K - red line, 1250 K - blue line, 1500 K - green line).

At higher temperatures, with increasing x in a particle density of electrons decreases rapidly. However, outside of the particle, the electron density is increases with increasing temperature.

In conclusion, we obtained the equation (4), which describes the distribution of the potential and the electron density in an equilibrium dust-electron plasma taking into account parameters of the electron gas inside the axial geometry dust particles. The inclusion of these parameters performed on the basis of the model of "solid-state plasma", considering the condensed particle system as the ion core and the free electron gas. Solving equation (4) we have been obtained the potential distributions and the electron density in thermal equilibrium dust-electron plasma. We have been found that near the particle surface there is a large gradient of the electric potential and as a result, the concentration of free electrons, i.e. electrons in the dust particle are located in a potential well. The obtained expressions allow to calculate the influence of several parameters such as temperature, concentration and dust particle size, type of the substance of particles on properties of the equilibrium dust-electron plasma.

Acknowledgements
The work is performed according to the Russian Government Program of Competitive Growth of Kazan Federal University.

References

[1] Complex and Dusty Plasmas, edited by V. E. Fortov and G. E. Morfill (CRC Press, Boca Raton, FL, 2010).
[2] V.I. Vishnyakov, Phy. Rev. E **85**, 026402 (2012).
[3] V.I. Vishnyakov and G.S. Dragan, Phys. Rev. E **73**, 026403 (2006).
[4] V.I. Vishnyakov, G.S. Dragan and V.M. Evtuhov, Phys. Rev. E **76**, 036402 (2007).
[5] Dautov G, Kashapov N, Larionov V, Fayrushin I et al. 2013 J. Phys.: Conf. Series **479** 011001
[6] Dautov G, Fayrushin I 2013 J. Phys.: Conf. Series **479** 012013
[7] Dautov G, Dautov I, Fayrushin I and Kashapov N 2013 J. Phys.: Conf. Series **479** 012014
[8] Dautov G, Dautov I, Fayrushin I and Kashapov N 2013 J. Phys.: Conf. Series **479** 012001
[9] R.D. Smirnov, A.Y. Pigarov, M. Rosenberg, S.I. Krasheninnikov, and D.A. Mendis, Plasma Phys. Controlled Fusion **49**, 347 (2007).
[10] J. Vaverka, I. Richterová, M. Vyšinka, J. Pavlü, J. Šafránková, and Z. Němeček, Plasma Phys. Controlled Fusion 56, 025001 (2014).
[11] V.I. Krauz, Yu.V. Martynenko, N.YuSvechnikov, V.P. Smirnov, V.G. Stankevich, L.N. Khimchenko, Phys. Usp.**53**1015–1038 (2010).
[12] W. Ekardt, Phys. Rev. B. 29, 1558 (1984).
[13] V.K. Ivanov, A.N. Ipatov, V.A. Kharchenko,JETP, Vol. 82, No 3, p. 485 (1996)
[14] M.B. Smirnov, V.P. Krainov, JETP, Vol. 88, No 6, p. 1102 (1999)
[15] N.W. Ashcroft and N. D. Mermin, Solid State Physics (Saunders College, Philadelphia, 1976).
[16] L.A. Artsimovich, R.Z. Sagdeev, Plasma physics for physicists. Moscow, Atomizdat, 1979. 320 p.
[17] B.M. Smirnov, Phys. Usp. **45** 1251–1286 (2002).
[18] L. D. Landau and E. M. Lifshitz, *Statisticheskaya Fizika*(Nauka, Moscow, 1976), Vol. 1 [*Statistical Physics* (Pergamon Press, Oxford, 1980)]

Journal of Physics: Conference Series **567** (2014) 012007 doi:10.1088/1742-6596/567/1/012007

Distribution of electric field near the surface of the aluminum oxide particle in the dust-electron thermal plasma

I Fayrushin, I Dautov and A Shamsutdinov
Department of Thechnical Physics and Power Engineering, Kazan Federal University, Kazan, 420008, Russia

E-mail: fairushin_ilnaz@mail.ru

Abstract. We obtained the equation, which describes the distribution of electrical field in an equilibrium dust-electron plasma taking into account parameters of the electron gas inside the dust particles. The inclusion of these parameters performed on the basis of the model of "solid-state plasma," considering the condensed particle system as the ion core and the free electron gas. These calculations are performed for aluminum oxide particles.

1. Introduction

The ionized gas containing particles of condensed substances, which are formed spontaneously or entered from the outside is called dust plasma. These particles are heated to a sufficiently high temperature, emitting electrons and positively charged, can significantly increase the degree of ionization of the plasma. Conversely, a "cold" particle may cause the recombination of free electrons [1, 2].

Dust-electron plasma is a system of dust particles, free electrons and neutral gas atoms, located at a temperature of 1000 to 3000 K and pressures of about atmospheric or above. Under these conditions the system is in a state of thermodynamic equilibrium is isothermal due to intensive exchange of heat between all components [3-7].

Processes occurring at the particle surface is largely determined by the value of the electric field in this area.

2. Theoretical background and results

Consider the potential distribution and the electron density in dusty plasmas on the basis of the model of "solid-state plasma," considering the condensed particle system as the ion core and the free electron gas [8-11]. We assume that the distance between two adjacent solid particles is equal to $2l$, solid particle radius R, its electron gas in the conduction band is a non-degenerate electron and ion densities are equal to n_e and n_i.

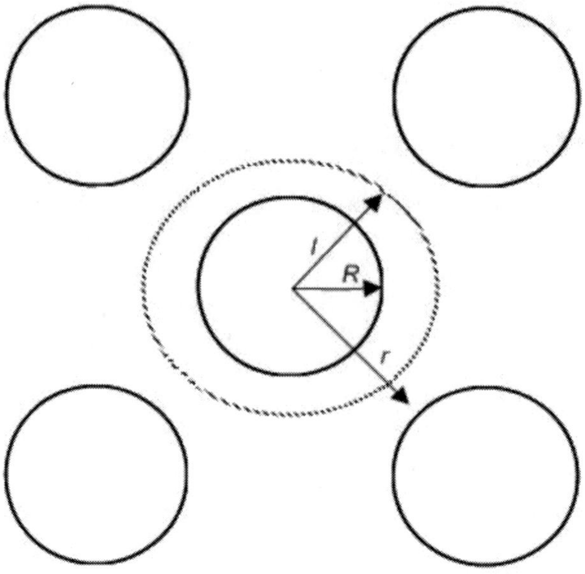

Fig. 1. Aa schematic representation of the dust particles in the dust-e plazme. R - particle radius, l - half the distance between the centers of adjacent chatits, r - relative distance to the center of the particle.

Gauss theorem for the region $0 \leq r \leq R$ written as

$$\oiint_S E dS = \frac{1}{\varepsilon\varepsilon_0} \iiint_V (n_i - n_e) q dV, \quad (1)$$

where ε - relative dielectric constant of the medium; ε_0 - electric constant; E - the projection of the electric field on the outer surface normal S, V - volume inside S; q - the absolute value of the electron charge. Using Boltzmann distribution for electrons

$$n_e = n_{e0} e^{\frac{q\varphi}{kT}}. \quad (2)$$

equation (1) for the case of spherically symmetric problem becomes Poisson-Boltzmann equation:

$$\frac{1}{r^2}\frac{d}{dr}\left(r^2 \frac{d\varphi}{dr}\right) = \frac{qn_{e0}}{\varepsilon\varepsilon_0}\left(e^{\frac{q\varphi}{kt}} - \frac{n_i}{n_{e_0}}\right), \quad (3)$$

where φ - the electric field potential, t - absolute temperature, k - Boltzmann constant.

By introducing the dimensionless quantities

$$x = \frac{r}{R}, \frac{q\varphi}{kt} = \psi, \overline{n}_i = \frac{n_i}{n_{e0}},$$

from equation (3) we obtain

$$\frac{1}{x^2}\frac{d}{dx}\left(x^2\frac{d\psi_1}{dx}\right)=a^2\left(e^{\Psi}-\bar{n}_i\right). \qquad (4)$$

Consider the case where $|\Psi_1|<1$. We expand the value in a row and only the first two members

$$e^{\psi_1}=1+\psi_1.$$

We introduce a value $\lambda=\frac{l}{R}$, wherein $2l$ - the average distance between the centers of two neighboring particles. In the $1\le x\le\lambda$ ions are absent, i.e. $\bar{n}_i=0$. With this in mind, from the expression (4) we obtain

$$\frac{1}{x^2}\frac{d}{dx}\left(x^2\frac{d\psi_1}{dx}\right)=a^2\left(1+\psi_1-\bar{n}_i\right), \qquad (5)$$

и

$$\frac{1}{x}\frac{d^2}{dx^2}\left(\psi_2 x\right)=a^2\left(1+\psi_2\right), \qquad (6)$$

where $a=qR\sqrt{\frac{n_{e0}}{kT\varepsilon\varepsilon_0}}$. With conditions $\psi_1(1)=\psi_2(1)=0,\ \psi_1'(1)=\psi_2'(1)$ и $\psi_2'(\lambda)=0$, solve equations (5) and (6). Obtain:

$$\psi_1(x)=\bar{n}_i-1+C\frac{\left(e^{ax}-e^{-ax}\right)}{x} \qquad (7)$$

$$\psi_2(x)=-1+\frac{D_1e^{ax}+D_2e^{-ax}}{x}, \qquad (8)$$

where

$$C=\frac{e^{2a(1-2\lambda)}\left(\frac{a\lambda+1}{a\lambda-1}\right)(a-1)-(a+1)}{\left(1+e^{2a(1-\lambda)}\frac{a\lambda+1}{a\lambda-1}\right)\cdot\left(e^{a}(a-1)+e^{-a}(a+1)\right)},$$

$$D_1=\frac{e^{a(1-2\lambda)}\frac{a\lambda+1}{a\lambda-1}}{1+e^{2a(1-\lambda)}\frac{a\lambda+1}{a\lambda-1}},$$

$$D_2=e^{a}-\frac{e^{a(1-2\lambda)}\frac{a\lambda+1}{a\lambda-1}}{1+e^{2a(1-\lambda)}\frac{a\lambda+1}{a\lambda-1}}e^{2a}.$$

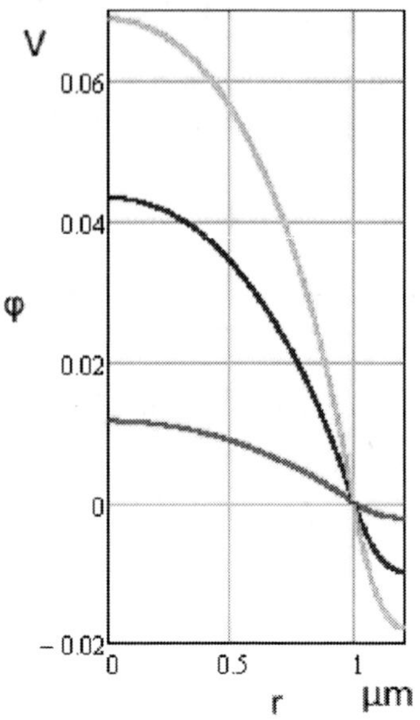

Fig. 2. The potential distribution at $R=10^{-6}$ м and different temperatures (1000 K - red line, 1100 K - blue line, 1200 K - green line)

As seen in Figure 2, the potential difference between the center of the particle and the surface of a sphere with a radius 1 and having a common center with the particle increases with increasing temperature. This is due to the increase in the degree of charge separation, due to higher electron density electron cloud.

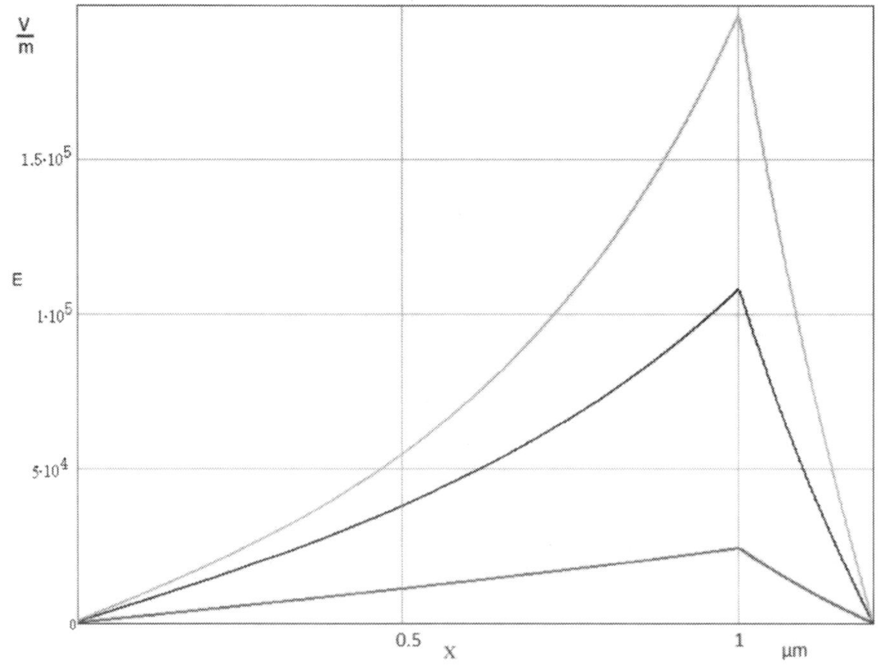

Fig. 3. The distribution of intensity in the region of the dust particles at different temperatures (1000 K - red line, 1100 K - blue line, 1200 K - green line)

From figure 3 it is seen that the field strength increases from the center of the particle and, reaching maximum values on its surface, drops sharply to the border region delineated by the distance l.

Acknowledgements
The work is performed according to the Russian Government Program of Competitive Growth of Kazan Federal University.

References
[1] Complex and Dusty Plasmas, edited by V. E. Fortov and G. E. Morfill (CRC Press, Boca Raton, FL, 2010).
[2] Fadeev S A and Kashapov N F 2013 J. Phys.: Conf. Series **479** 012009
[3] V.I. Vishnyakov, Phy. Rev. E **85**, 026402 (2012).
[4] Dautov G, Kashapov N, Larionov V, Fayrushin I et al. 2013 J. Phys.: Conf. Series **479** 011001
[5] Dautov G, Fayrushin I 2013 J. Phys.: Conf. Series **479** 012013
[6] Dautov G, Dautov I, Fayrushin I and Kashapov N 2013 J. Phys.: Conf. Series **479** 012014
[7] Dautov G, Dautov I, Fayrushin I and Kashapov N 2013 J. Phys.: Conf. Series **479** 012001
[8] N.W. Ashcroft and N. D. Mermin, Solid State Physics (Saunders College, Philadelphia, 1976).
[9] L.A. Artsimovich, R.Z. Sagdeev, Plasma physics for physicists. Moscow, Atomizdat, 1979 p 320
[10] B.M. Smirnov, Phys. Usp. **45** 1251–1286 (2002).
[11] L. D. Landau and E. M. Lifshitz, *Statisticheskaya Fizika* (Nauka, Moscow, 1976), Vol. 1 [*Statistical Physics* (Pergamon Press, Oxford, 1980)]

Journal of Physics: Conference Series **567** (2014) 012008 doi:10.1088/1742-6596/567/1/012008

Calculation of the surface tension energy for the metal dust particle in the dust-electron thermal plasma

G Dautov[1], I Fayrushin[2], N Kashapov[2], and I Dautov[2]
[1]General Physics Department, Kazan National Research Technical University by named A.N. Tupolev-KAI, Kazan, 420111, Russia
[2]Department of Thechnical Physics and Power Engineering, Kazan Federal University, Kazan, 420008, Russia

E-mail: fairushin_ilnaz@mail.ru

Abstract. In this work, we calculated the energy values for the surface tension of the dust particles in the plasma on the basis of the analytical solution of the Poisson-Fermi-Dirac ecuatiins. Calculations showed an increase of the surface tension energy with decreasing particle radius.

1. Introduction

Processes to obtain plasma coatings from powder materials involves heating and accelerating particles in the flow of the thermal plasma[1-11]. The issue of energy dependence of the surface tension of the dust particles is very relevant, because determines the quality of the coatings.

To determine these characteristics of the data which are used in this paper the model of "jelly", which previously has been widely used in the description of the electrical properties of atomic clusters [12-14]. According to this model, the dust particle is represented as two-components system. The first component is ion core, which create uniform positive background in the whole volume of the dust particle. The second component is the electronic gas. Thus, in this model, a dust particle can be regarded as the area of solid state plasma [15-17].

2. Theoretical background and results

It is known from the statistical physics, the properties of the electron gas are described by the Fermi-Dirac distribution

$$n_j = \frac{1}{1+e^{\frac{\varepsilon_j-\mu}{\theta}}}, \tag{1}$$

where θ – statistical temperature equal kT, k - Boltzmann constant, T- absolute temperature, μ – Fermi energy, ε_j - electron energy in the j -th quantum state, n_j - average numbers of electrons in the same quantum state with energy of ε_j.

In the semiclassical approximation, the number of electrons per unit volume in the energy range from ε to $\varepsilon + d\varepsilon$ defined by the formula [18]

$$n_e(\varepsilon)=\int_0^{\infty}\frac{4\pi\cdot(2m_e)^{\frac{3}{2}}\varepsilon^{\frac{1}{2}}}{h^3\left(1+e^{\frac{\varepsilon-\mu}{\theta}}\right)}\cdot d\varepsilon\,. \tag{2}$$

Here h – Planck's constant, m_e - mass of the electron, n_e - electron density.

When $\theta = 0$, the electron gas is completely degenerate and the equation (2) is obtained from the Fermi energy:

$$\mu=\frac{h^2}{2m_e}\cdot\left(\frac{3n_e}{8\pi}\right)^{\frac{2}{3}}. \tag{3}$$

The distribution of the potential φ of the electrostatic field is described by Poisson equation

$$\Delta\varphi=\frac{n_e-n_i}{\varepsilon_0\varepsilon_1}q\,, \tag{5}$$

where ε_1 - relative permittivity, ε_0 - electric constant, n_i - the concentration of positive charge of the ion core (the density of the positive background), q - the absolute value of the electron charge.

The condition of statistical equilibrium of the electron gas can be written as[14]

$$\mu-q\varphi=const\,. \tag{6}$$

Equations (2), (5) and (6) give the boundary conditions completely determine the spatial distribution of the potential and the electron density.

The system of equations (2), (5) and (6) can be obtained by a single equation, which calculates the potential distribution and electron density. From (6) wefind

$$\Delta\varphi=\frac{\Delta\mu}{q}. \tag{7}$$

Substitution of expressions (2) and (7) into (5) leads to a nonlinear integro-differential equation

$$\Delta\mu=\frac{q^2}{\varepsilon_0\varepsilon_1}\left[\int_0^{\infty}\frac{4\pi(2m_e)^{\frac{3}{2}}\varepsilon^{\frac{1}{2}}}{h^3\left(1+e^{\frac{\varepsilon-\mu}{\theta}}\right)}d\varepsilon-n_i\right]. \tag{8}$$

This equation is applicable to both degenerate and non-degenerate electron gas. This is particularly important in the case of dust particles from the metal. In this case, there are areas of the degenerate and non-degenerate electron gas.

In the case of degenerationgas (2) we find

$$n_e = \frac{16\sqrt{2}\pi m^{3/2}}{3h^3}\mu^{3/2}. \tag{9}$$

Substituting this expression in (8), we obtain

$$\Delta\mu = \frac{q^2}{\varepsilon_0\varepsilon_1}\left(\frac{16\sqrt{2}\pi m^{3/2}}{3h^3}\mu^{3/2} - n_i\right). \tag{10}$$

If we introduce the dimensionless potential $\psi = \frac{q\varphi}{\mu_0}$, for the case of spherical symmetry from the (7) and (10) we obtain the Poisson-Fermi-Dirac equations

$$\frac{1}{x^2}\cdot\frac{d}{dx}\left(x^2\cdot\frac{d\psi}{dx}\right) = a^2\left[(1+\psi)^{3/2} - \bar{n}_i\right], \tag{11}$$

where $x = \frac{r}{R}$, $\frac{n_i}{n_{e0}} = \bar{n}_i$, $a^2 = \frac{q^2R^2n_{e0}}{\mu_0\varepsilon_1\varepsilon_0}$.

Consider the case, when $|\psi| < 1$. We expand the value $(1+\psi)^{3/2}$ in the series and retain the first two terms: $(1+\psi)^{3/2} = 1 + \frac{3}{2}\psi$.

Given of this expression of the equation(11) obtained by

$$\frac{1}{x^2}\cdot\frac{d}{dx}\left(x^2\cdot\frac{d\psi}{dx}\right) = a^2\left(1 + \frac{3}{2}\psi - \bar{n}_i\right). \tag{12}$$

The initial conditions for this equation is $\psi(0) = 0$, $\psi'(0) = 0$. Its solution is

$$\psi(x) = \frac{2}{3}\cdot\frac{(a-1)\cdot e^a - \left(\frac{a\lambda-1}{a\lambda+1}\right)\cdot(a+1)\cdot e^{a(2\lambda-1)}}{2a\cdot\left[(a-1)\cdot e^a - (a+1)\cdot\left(\frac{a\lambda-1}{a\lambda+1}\right)\cdot e^{a(2\lambda-1)} + \left(\frac{a\lambda-1}{a\lambda+1}\right)\cdot e^{2a\lambda} + 1\right]}\cdot\left[\frac{\left(e^{ax} - e^{-ax}\right)}{x} - 2a\right], \text{if } 0 \le x \le 1 \tag{13}$$

If the concentration of dust particles is large; concentration of electrons in the region $1 < x \le \lambda$ will be large and the electron gas is degenerate. In this case, the potential distribution in the area $1 < x \le \lambda$ defined by the formula.

$$\psi(x) = \frac{2}{3}\cdot\left(\frac{(a-1)\cdot e^a + (a+1)\cdot e^{-a}}{2a\cdot\left((a-1)\cdot e^a - (a+1)\cdot\left(\frac{a\lambda-1}{a\lambda+1}\right)\cdot e^{a(2\lambda-1)} + \left(\frac{a\lambda-1}{a\lambda+1}\right)\cdot e^{2a\lambda} + 1\right)}\cdot\frac{e^{ax} + \left(\frac{a\lambda-1}{a\lambda+1}\right)\cdot e^{a(2\lambda-x)}}{x} - 1\right), \text{ if } 1 < x \le \lambda \tag{14}$$

Surface tension energy is the total electrostatic energy W in double layer near dust particle per square of surface of particle S [19]

$$\sigma = \frac{W}{S} = \frac{\varepsilon_1 \varepsilon_0}{2R^2} \int_0^l (E(r))^2 \cdot r^2 dr, \tag{15}$$

where $E(r) = -\frac{d\varphi}{dr}$ - electrical field intensity. Calculations showed an increase of the surface tension energy with decreasing particle radius.

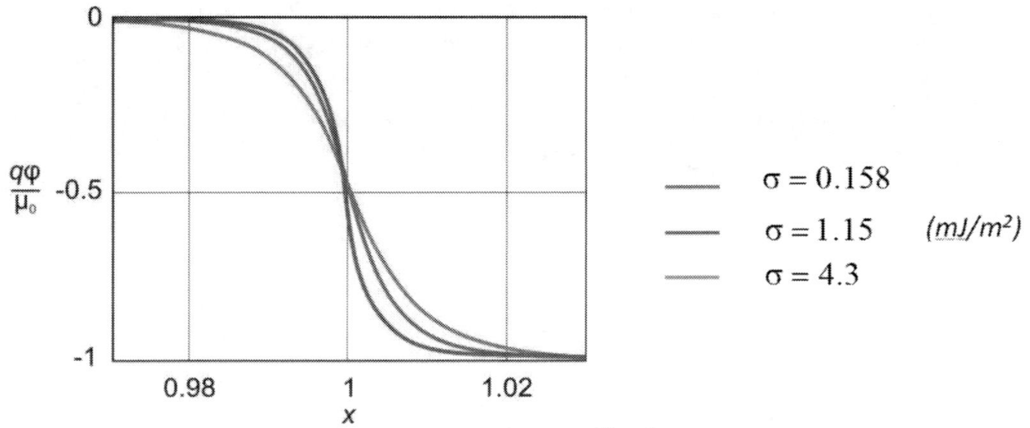

Fig. Distribution the dimesionless potential at $n_i = 10^{27}$ m^{-3}and different of the radius of the dust particle ($R = 2 \cdot 10^{-8}$ m – green line, $R = 3 \cdot 10^{-8}$ m – blue line, $R = 4 \cdot 10^{-8}$ m – red line) and surface tension energy σ.

Acknowledgements
The work is performed according to the Russian Government Program of Competitive Growth of Kazan Federal University.

References

[1] Complex and Dusty Plasmas, edited by V. E. Fortov and G. E. Morfill (CRC Press, Boca Raton, FL, 2010).
[2] V.I. Vishnyakov, Phy. Rev. E **85**, 026402 (2012).
[3] V.I. Vishnyakov and G.S. Dragan, Phys. Rev. E **73**, 026403 (2006).
[4] V.I. Vishnyakov, G.S. Dragan and V.M. Evtuhov, Phys. Rev. E **76**, 036402 (2007).
[5] Dautov G, Kashapov N, Larionov V, Fayrushin I et al. 2013 J. Phys.: Conf. Series **479** 011001
[6] Dautov G, Fayrushin I 2013 J. Phys.: Conf. Series **479** 012013
[7] Dautov G, Dautov I, Fayrushin I and Kashapov N 2013 J. Phys.: Conf. Series **479** 012014
[8] Dautov G, Dautov I, Fayrushin I and Kashapov N 2013 J. Phys.: Conf. Series **479** 012001
[9] R.D. Smirnov, A.Y. Pigarov, M. Rosenberg, S.I. Krasheninnikov, and D.A. Mendis, Plasma Phys. Controlled Fusion **49**, 347 (2007).
[10] J. Vaverka, I. Richterová, M. Vyšinka, J. Pavlü, J. Šafránková, and Z. Němeček, Plasma Phys. Controlled Fusion 56, 025001 (2014).

[11] V.I. Krauz, Yu.V. Martynenko, N.YuSvechnikov, V.P. Smirnov, V.G. Stankevich, L.N. Khimchenko, Phys. Usp.**53**1015–1038 (2010).
[12] W. Ekardt, Phys. Rev. B. 29, 1558 (1984).
[13] V.K. Ivanov, A.N. Ipatov, V.A. Kharchenko,JETP, Vol. 82, No 3, p. 485 (1996)
[14] M.B. Smirnov, V.P. Krainov, JETP, Vol. 88, No 6, p. 1102 (1999)
[15] N.W. Ashcroft and N. D. Mermin, Solid State Physics (Saunders College, Philadelphia, 1976).
[16] L.A. Artsimovich, R.Z. Sagdeev, Plasma physics for physicists. Moscow, Atomizdat, 1979. 320 p.
[17] B.M. Smirnov, Phys. Usp. **45** 1251–1286 (2002).
[18] L. D. Landau and E. M. Lifshitz, *Statisticheskaya Fizika*(Nauka, Moscow, 1976), Vol. 1 [*Statistical Physics* (Pergamon Press, Oxford, 1980)]
[19] V. I. Roldugin, *Fizikohimiya poverhnosti* (Intellect, Dolgoprudny, 2011), 568 p.

Journal of Physics: Conference Series **567** (2014) 012009 doi:10.1088/1742-6596/567/1/012009

Influence of metal macroparticles to the electron density in a dusty plasma

I Fayrushin, N Kashapov, and I Dautov
[2]Department of Thechnical Physics and Power Engineering, Kazan Federal University, Kazan, 420008, Russia

E-mail: fairushin_ilnaz@mail.ru

Abstract. We obtained the equation, which describes electron density in an equilibrium dusty plasma taking into account parameters of the electron gas inside the dust particles. The inclusion of these parameters performed on the basis of the model of "solid-state plasma," considering the condensed particle system as the ion core and the free electron gas. The shape of potential barrier at the particle surface assumed as a rectangular.

1. Introduction

Dust particles in a plasma acquire an electric charge and are an additional charged component of the plasma. The properties of dusty plasmas are much richer than those of ordinary multi-component plasma of electrons and ions of various sorts [1]. In technological processes of plasma and flame spraying the powder particles are entered into the hot gas stream. Heated particles emit electrons into the surrounding gas or plasma and acquire a positive charge forming dusty plasma as a result [2-8]. The presence of charged dust particles has a strong influence on the characteristics of the flow. Intensive work is being done in the physics of dusty gas clouds in the atmosphere. The presence of charged dusty particles in the lower ionosphere radically affect its ionization properties.

In connection with the above description experimental and theoretical studies of the dusty plasmas properties are of great interest.

We consider dusty plasma at low temperatures, where thermal ionization of the gas can be neglected.

2. Theoretical background and results

We assume that the macroparticles are spheres of radius R, their concentration is equal to n_r, the concentration of electrons in the conduction band of dusty particles before emission is equal to n_{e0}, the average concentration of electrons in a gas or plasma, n_e, one dusty particle emits N electrons, the electron gas and dusty particles are in a state of statistical of equilibrium [9-12].

The following equation [12] equilibrium condition of the electron gas

$$\mu_1 + e\varphi_1 = \mu + e\varphi, \tag{1}$$

where μ_1 - the Fermi energy for an electron within the dusty particle and μ - is the Fermi energy for an electron outside of the dusty particle, φ_1 and φ - potentials of the electric field inside and outside of the dusty particle, e - charge of an electron. We assume that the $\varphi=0$ and inside of the dusty particle $\varphi_1=const$, $|e|=q$. With this in mind, equation (1) can be written as

$$\mu_1-\mu=q\varphi_1 . \tag{2}$$

Here h – Planck's constant, m_e - mass of the electron, n_e - electron density.

The potential energy of an electron within the dusty particle with a charge equal to Nq, is determined by the known equation

$$W_n=-q\varphi_1=-\frac{q^2N}{4\pi\varepsilon\varepsilon_0 R} . \tag{3}$$

where ε - relative dielectric permittivity, ε_0 - electric constant.

The concentration of free electrons in a gas or plasma is usually low, and therefore as μ it necessary to use the Fermi energy of a nondegenerate electron gas

$$\mu=\theta\ln(an_e),\ a=\frac{1}{2}\left(\frac{h^2}{2\pi m_e\theta}\right)^{\frac{3}{2}} . \tag{4}$$

Here, h - Planck constant, θ - statistical temperature, equal to kT, k - Boltzmann constant, T - absolute temperature, m_e – mass of the electron.

Depending on temperature and concentration of electrons in dusty particles are two possibilities. For large n_{e0} and small T the electronic gas in a dusty particle is degenerate, while for small n_{e0} and large T it is nondegenerated. Lets consider first case in more detail. The Fermi energy of the electronic gas in a dusty particle is defined by

$$\mu_1=\frac{h^2}{2m_e}\left(\frac{3n_{e1}}{8\pi}\right)^{\frac{2}{3}} . \tag{5}$$

where n_{e1} - concentration of the electrons in a dusty particle after emission.

Substituting expressions (3), (4) and (5) in (2) gives

$$\frac{h^2}{2m_e}\left(\frac{3n_{e1}}{8\pi}\right)^{\frac{2}{3}}-\theta\ln(an_e)=bN,\ b=\frac{q^2}{4\pi\varepsilon\varepsilon_0 R} . \tag{6}$$

Due to the absence of the thermal ionization of the gas we have

$$n_r N=n_e . \tag{7}$$

From (6) we find

$$\frac{h^2}{2m_e}\left(\frac{3\left(n_{e0}-\frac{3N}{4\pi R^3}\right)}{8\pi}\right)^{\frac{2}{3}}-\theta\ln(an_r N)=bN . \tag{8}$$

The figure shows the dependence of n_e from R when $\varepsilon = 1$, $n_{e0} = 10^{24}$ m^{-3}, and different values of temperature (curve 1 - T = 1000 K, curve 2 - T = 1500 K, curve 3 - T = 2000 K). As can be seen, a decrease of R and increase of T leads to an increase of the electrons concentration in a dusty plasma.

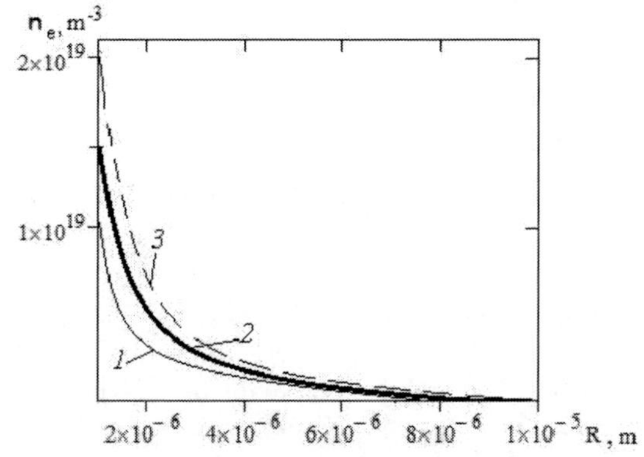

Fig.

Acknowledgements
The work is performed according to the Russian Government Program of Competitive Growth of Kazan Federal University.

References

[1] Complex and Dusty Plasmas, edited by V. E. Fortov and G. E. Morfill (CRC Press, Boca Raton, FL, 2010).
[2] V.I. Vishnyakov, Phy. Rev. E **85**, 026402 (2012).
[3] V.I. Vishnyakov and G.S. Dragan, Phys. Rev. E **73**, 026403 (2006).
[4] V.I. Vishnyakov, G.S. Dragan and V.M. Evtuhov, Phys. Rev. E **76**, 036402 (2007).
[5] Dautov G, Kashapov N, Larionov V, Fayrushin I et al. 2013 J. Phys.: Conf. Series **479** 011001
[6] Dautov G, Fayrushin I 2013 J. Phys.: Conf. Series **479** 012013
[7] Dautov G, Dautov I, Fayrushin I and Kashapov N 2013 J. Phys.: Conf. Series **479** 012014
[8] Dautov G, Dautov I, Fayrushin I and Kashapov N 2013 J. Phys.: Conf. Series **479** 012001
[9] M.B. Smirnov, V.P. Krainov, JETP, Vol. 88, No 6, p. 1102 (1999)
[10] N.W. Ashcroft and N. D. Mermin, Solid State Physics (Saunders College, Philadelphia, 1976).
[11] L.A. Artsimovich, R.Z. Sagdeev, Plasma physics for physicists. Moscow, Atomizdat, 1979. 320 p.
[12] L. D. Landau and E. M. Lifshitz, *Statisticheskaya Fizika*(Nauka, Moscow, 1976), Vol. 1 [*Statistical Physics* (Pergamon Press, Oxford, 1980)]

Journal of Physics: Conference Series **567** (2014) 012010 doi:10.1088/1742-6596/567/1/012010

Air plasma sprayed coatings of self-fluxing powder materials

E.E. Kornienko[1], E.J. Lapushkina[1], V.I. Kuzmin[2], S.P. Vaschenko[2], I.P. Gulyaev[2], E.V. Kartaev[2], D.S. Sergachev[2], N Kashapov[3], S Sharifullin[3], I Fayrushin[3]
[1]Novosibirsky State Technical University, Novosibirsk, Russia
[2]Institute of Theoretical and Applied Mechanics SB RAS, Novosibirsk, Russia
[3]Department of Thechnical Physics and Power Engineering, Kazan Federal University, Kazan, 420008, Russia

E-mail: [1]kornienko_ee@mail.ru, [2]vikuzmin57@mail.ru, [3]Saidchist@mail.ru

Abstract. The article discusses the structural features of self-fluxing coatings obtained by plasma spraying air from entering the hub ring and the gas-dynamic focusing powder. It was shown that, unlike the unilateral spot powder inlet into the plasma jet, the use of the annular input node allows to increase heating efficiency and to accelerate the particles in the plasma stream. By optical and scanning electron microscopy that most of the particles forming the coating, in the plasma jet is in a molten or plasticized condition. Transmission electron microscopy revealed that high cooling rates of such particles contribute to the formation of γ-SMC supersaturated solid solution Ni-based average grain size of 80 nm.

1. Introduction

One of the most urgent problems of modern materials science is the development of new structural materials with high technological properties and can withstand increasingly tough operating conditions. Depending on the purpose of product materials from which they are made should possess various properties. For example a high wear resistance, heat resistance, thermal conductivity, strength, corrosion resistance, electrical conductivity, etc.

Operation of machine parts is accompanied by deterioration, which in turn requires replacement and increases production costs. Furthermore, to make the entire piece of expensive material is impractical because it is related to the high cost of new materials and the difficulty of processing the workpiece surface. Effective solution to the problem of increasing the wear resistance of the surface is the formation of coatings on steel products [1]. Furthermore, coating - a way to save scarce and expensive alloying elements which are used to produce steels and alloys with special properties [1 - 3].

To improve the durability of parts and increase its service life can use various methods of surface hardening. One of the promising methods of coating formation is currently air-plasma spraying powder materials [1-6]. This technology allows you to apply coatings of different materials on the surface of any complexity or restore the most worn surfaces of the parts [2, 4].

The most common way to enter the powder into the plasma jet cross point is administered through a tube-shihtoprovod shear outlet nozzle or through a hole in the nozzle [7-9]. This method contributes to heterogeneity of the input fields of temperature and velocity of the plasma jet, which considerably reduces the quality of the formed coatings. The coatings are characterized by high porosity (15%) and poor adhesion [10, 11], which prevents their widespread use.

For the axisymmetric high-temperature heterogeneous flow in the present study used the input node of the ring of powder with gas-dynamic focusing. The prospect of such a powder inlet and the existence of the above problems have led to the development of the original node ring input and relevant research [12].

In contrast to the one-sided point of input feed conveyor ring and focusing gases does not lead to a distortion of the temperature profiles of the plasma jet, as can be seen in Figure 1, which shows the temperature profiles of air plasma jet at a distance of 5 mm from the nozzle as a free flow conveyor and focusing gases (curve 1) and in their feed (curve 2) [13].

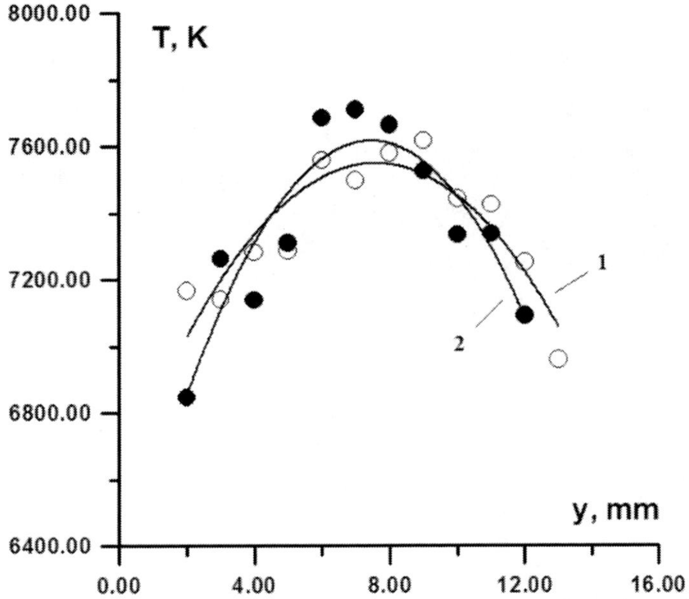

Fig.1. The temperature distribution in the plasma jet at a distance z = 5mm from the edge of the outlet nozzle at the operating conditions of the plasma torch: I = 200 A, U = 270 B, Gplazm. = 2 × 10-3kg / s; Gzav = 0.25 × 10-3 kg / s; 1 - GTP = Gfok = 0 kg / s; 2 - GTP = Gfok = 0,5 × 10-3 kg / s

Measurements of the temperature distribution in the plasma jet formed by the spectral lines of atomic copper Cu1, present in the air stream due to erosion of the copper electrode (anode). The spectra were measured for several positions line up. Scanning the diameter of the plasma jet was conducted by mechanical movement of the plasma torch 1 mm increments from the bottom up. Position y = 0 corresponds to the uppermost position of the jet, y = 14 mm - the lowest position of the jet, in which the visible spectral lines. The spectrum contains lines Cu1 λ = 5105.6 Å, 5153.2 Å, 5218.2 Å, 5700.2 Å and 5782.1 Å. To measure the temperature of the plasma were selected pairs of lines: λ1 = 5105.6 Å, λ2 = 5153.2 Å; λ1 = 5105.6 Å, λ3 = 5218.2 Å.

As seen from the graph in Figure 1, a slight decrease in temperature, feeding the carrier gas and the focusing is observed only in the periphery of the jet. The central part of the jet remains undisturbed, indicating an insignificant depth of penetration into the plasma jet of cold carrier gas and a focusing assembly using annular input powdered materials.

Earlier work by the authors showed that use of an annular input node powder, compared with a one-way point, allows to significantly increase the heating efficiency of the particles, which contributes significantly to the quality of ceramic and metallic coatings [13 - 19].

In this paper, as a material for the deposition of wear-resistant coatings of self-fluxing powder was selected on the basis of nickel. Due to the high resistance of the formed coating to abrasion and ability to work with significant impact loads, it is used in machine tool (protection guides machines). High heat resistance causes the application of self-fluxing coatings in aircraft (turbine blades and compressor of a jet engine) and mechanical engineering (protection detachable and permanent matrices for pressing). The high corrosion resistance allows their use in the petroleum and coal industry (production of parts of mud pumps). High resistance to cavitation wear allows the use of these coatings in transport engineering (propellers river and sea vessels) [1-3, 17].

The aim of this work was to study the structure and properties of wear-resistant coatings of self-fluxing alloys formed by plasma spraying using the input node of the ring of powder with gas-dynamic focusing.

2. Materials and Methods of Research

As the material for the formation of wear-resistant coatings used self-fluxing nickel-based powder brand PR N77H15S3R2-3 dispersion 40/100 microns. The powder was sprayed onto a mild steel pipe St3 with 25 mm inside diameter and a wall thickness of 3 mm.

Plasma spraying was performed using a plasma torch PNA-50 manned annular input node of powder with gas-dynamic focusing (development of the Institute of Theoretical and Applied Mechanics SB RAS). Spraying was carried out on samples of the following modes:

Mode number 1: current - 140 A, the voltage - 265 V;
mode number 2: current - 170 A, the voltage - 258 V;
mode number 3: current - 200 A, the voltage - 250 V;
mode number 4: current - 230 A, the voltage of - 243 V.

Spraying distance was 170 mm. As the plasma gas used air as a protective (veil anode), and the focusing of the conveying gas, a mixture of air and propane-butane. Plasma gas flow was 3 g / s, shielding - 0.45 g / s, and the conveying focusing - 0.9 g / s.

Immediately before spraying, blasting the surface of the parent metal. Measuring temperature of heating the particles in a plasma jet were performed using the apparatus and methods presented in [20].

For metallographic studies using optical microscope Carl Zeiss Axio Observer A1m. Deeper structural studies were carried out on a scanning electron microscope Carl Zeiss EVO50 XVP using microprobe EDS X-Act and the transmission electron microscope Tecnai G2 FEI. To determine the phase composition of the coatings used X-ray diffractometer ARL X'TRA. The porosity of the coatings was determined by microscopic method. The microhardness of the structural components of coatings were evaluated for Wolpert Group 402MVD. Wear-resistant coatings in friction of loosely fixed abrasive particles was evaluated by comparing the wear test samples and specimens of mild steel 20 after cementation according to GOST 23.208-79 [21].

3. Results and discussion

In [15] it was shown that the formed coatings are characterized by a complex structure: there are not melted or deformed (the first type); not melted but deform plastically (second type) and molten particles (the third type). The number of particles of the first type in the coating is large, and with increasing of the arc current remains practically unchanged. This caused a slight increase in the temperature of particles in the plasma jet. Table. 1 shows the mean temperature of the spray particles in the cross section of the plasma jet at a distance of 170 mm from the nozzle.

Table 1.

Dependence of the average temperature of the powder particles of the PR-N77H15S3R2-3 current arc plasma torch

deposition mode	Average temperature, K
number 1	2189 ± 199

number 2	2249 ± 220
number 3	2292 ± 236
number 4	2383 ± 250

As a result, local microhardness measurements revealed that the maximum hardness (900 ... 1050 HV) have a third type of particle. This is due to the structural changes occurring in the particles of this type during the crystallization. Firstly, heating of the particles within the plasma stream resulting in the formation of a homogeneous melt, and rapid crystallization (solidification time of the particles - 10-5 ... 10-7 [1]) to the fixing metastable structures (supersaturated solid solution), which is typical during deposition of alloys a quantity of alloying elements which exceeds the limit of their solubility in the solid state. According to X-ray structure analysis in the presence of the coating was found γ-based solid solution Ni. Presumably, this supersaturated solid solution of Cr and Si in the γ-Ni. Furthermore, X-ray diffraction analysis were able to detect the presence in the coating of the reinforcing phase in the form of nickel boride (Ni_3B), chromium boride (Cr_2B and CrB), chromium carbide (Cr_7C_3), nickel silicide (Ni_3Si) and nickel-iron ($FeNi_3$).

Transmission electron microscopy revealed that the amount of discharge of the hardening phase is from 5 nm to 50 nm. Since sputtering was performed in air, in the structure of the coatings were also found nickel nitride (Ni_2N) and nickel oxide (Ni_2O). Furthermore, it was found that the particles of the third type is formed submicrocrystalline structure with an average grain size of 80 nm.

One disadvantage of plasma spray technology with a point entering the powder coating is a high porosity (15%). Porosity of coatings formed by means of plasma spraying with an annular entry, far below (Table. 2). This phenomenon can be explained by the fact that entering the annular powder provides a dense axisymmetric heterogeneous flow, which in turn contributes to the maximum flow rate of the plasma and the maximum penetration of the powder particles.

Significant differences between the formation of heterogeneous flows in the ring and dot powder inlet are clearly visible in the photographs presented in Figure 2 [13].

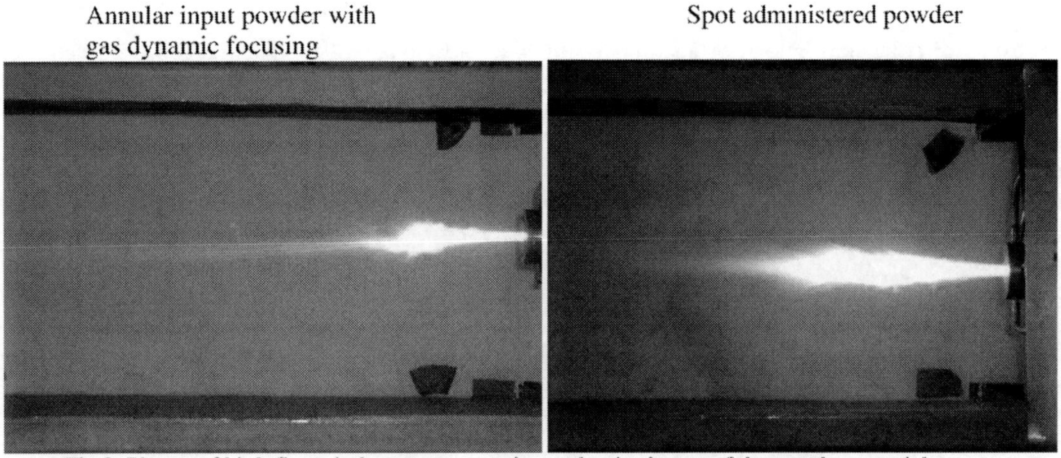

Fig.2. Photos of high flows in heterogeneous ring and point inputs of the powder material

It is clearly seen that, in contrast to a point input, the input ring with gas dynamic focusing provides dense axisymmetric heterogeneous stream (with the correct mode, all the powder particles pass through a high-temperature and high-axial region of the plasma jet).

Coatings formed on mode number 1 have the lowest porosity (0.96%), in contrast to the coatings obtained on the mode number 4 (~ 5%). This phenomenon can be explained by the increase in the degree of superheat of the particles. As noted above, the majority of sputtered particles are melted or plasticized (particles of a first or second type). Subsequent crystallization of the particles occurs with a decrease in volume, which contributes to pore within the particle, and on the interfaces therebetween.

Table 2.

Pore size distribution depending on the current strength

Number mode	Porosity,%	Pore size, microns			
		< 5	5 – 10	11 – 20	≥ 30
		The pore distribution on the coating,%			
1	0,96	52	28	20	0
2	1,95	29	29	36	5
3	3,11	18	28	44	10
4	4,95	21	22	37	20

The pore size distribution according to the mode of plasma spraying are presented in Table 2. In the coatings formed on the mode number 1, more than 50% of the pores have a size less than 5 microns. Such pores often occur either within the unmelted particles (defect inherent to the powder as well in the initial state) or at the interfaces between particles. Pore size greater than 30 .mu.m is observed in contrast to coatings produced at 230 A (20%).

Note that the wrong mode selected deposition may contribute unstuck. High speed cooling of the molten particles contribute to the formation of tensile residual stresses in the coatings. When the residual stresses exceed the strength of adhesion, coating peels off spontaneously. As follows from the analysis of the structures of the transition zone "basecoats" adhesion of the coating obtained on the mode number 1 (less overheating particles) is high enough. Increasing the arc current contributes to higher residual stresses in the coatings.

The results of comparative tests of coatings on the wear resistance of the friction loosely fixed abrasive particles showed that the tribological properties of the coatings significantly reduced with increasing current arc plasma torch, which correlates well with the results of measurements of porosity. Wear resistance of the coatings formed at 70% higher wear resistance of specimens of steel 20 after cementation.

Conclusions

1. The plasma spray technology using the input node of the ring of powder with gas-dynamic focusing allows you to create high-quality coatings with low porosity, high wear resistance, hardness and adhesion.
2. High speed cooling of the particles on the substrate contribute to the formation of γ-SMC supersaturated solid solution Ni-based average grain size of 80 nm. Furthermore, in the coating are allocated reinforcing phase nickel boride (Ni_3B), chromium boride (Cr_2B and CrB), chromium carbide (Cr_7C_3), nickel silicide (Ni_3Si), nickel-iron ($FeNi_3$), the dimensions of which are from 5 to 50 nm.
3. The tribological properties of the coatings slightly decreased with increasing arc current, which correlates well with the results of measurements of porosity. Wear resistance of the coatings formed at 70% higher wear resistance of specimens of steel 20 after cementation.

Acknowledgements

[1]This work was financially supported by the Ministry of Education and Science of the Russian Federation within the framework of the Strategic Development Program NSTU (Project C-5).

[3] The work is performed according to the Russian Government Program of Competitive Growth of Kazan Federal University.

References

[1] M. Hocking, Vasantasri V., P. Sidky metallic and ceramic coatings. Preparation, properties and applications. Per. from English. - M .: Mir, 2000. - 518 p.

[2] A. Hasuy, Morigaki O. surfacing and spraying. M .: Mechanical Engineering, 1985. 240 p.

[3] LI Tushino, bad AV, AO Tokarev, VI Sindeev Methods of research materials: structure, properties and processes of deposition of inorganic coatings. - M .: Mir, 2004. - 384 p.

[4] Ilyushchenko AF, AI Shevtsov, VA Okovity and others. The formation of thermal coatings and their modeling. - Minsk: Belarus. Navuka, 2011. - 357 p. - ISBN 978-985-08-1270-4.

[5] Dautov G, Fayrushin I 2013 J. Phys.: Conf. Series **479** 012013

[6] Dautov G, Dautov I, Fayrushin I and Kashapov N 2013 J. Phys.: Conf. Series **479** 012014

[7] P. Fauchais and M. Vardelle, Sensors in Spray Processes // J. Thermal Spray Technology, 2010, 19 (4), p 668 - 694.

[8] L. An, Y. Gao, and T. Zhang, Effect of Powder Injection Location on Ceramic Coatings Properties When Using Plasma Spray, J. Thermal Spray Technology, 2007, 16 (5-6), p 967 - 973.

[9] G. Mauer, R. Vassen, D. Stover and others, Improving Powder Injection in Plasma Spraying by Optical Diagnostics of the Plasma and Particle Characterization, J. Thermal Spray Technology, 2011, 20 (1-2), p 3 - 11.

[10] AF Puzryakov Theoretical Foundations of plasma spraying. - M .: Publishing House of the MSTU. NE Bauman, 2003. - 360 p.

[11] Chësov YS, EA Zverev The influence of technological factors on the properties of plasma spraying powder coatings brand PG-C27 // Metal. - Novosibirsk: Publishing House of the NSTU. -2009. - № 3 (44). - S. 15 - 17.

[12] VI Kuzmin, AA Mikhalchanka, EV Cartan. Input node of the ring of powder materials arc plasma torch. RF Patent №2474983. BIMP №4, 10.02.2013, the

[13] V.I. Kuzmin,, A.A. Mikhal'chenko, O.B. Kovalev, E.V. Kartaev, N.A. Rudenskaya. The technique of formation of the axisymmetric heterogeneous flow for thermal spraying of powder materials // J. of Thermal Spray Technology, 2012, V. 21, No. 1, p.159 - 168. DOI: 10.1007 / s11666-011-9701-6

[14] Kornienko EE, Nikulina AA, Lapushkina E.U. Structural research of boron carbide coatings received by plasma spraying. International conference on surface engineering for research and industrial applications, (Interfinish - Seria 2014), Russia Novosibirsk, June 30 - July 4, 2014. - P. 68.

[15] EE Kornienko, Lapushkina EY, Sergachev DV Plasma spraying of wear-resistant coatings. Modern equipment and technologies: Sat. Dokl. 20 int. scientific and practical. Conf. graduate students and young scientists, Tomsk, 14-18 apr.2014 In 3 vol. - Tomsk: TPU, 2014. - T. 2. - P. 51 - 52 - 100 copies.

[16] V. Kuzmin, E. Cartan, Kornienko EE et al. Plasma spraying of powder coatings with gas dynamic focusing of the dispersed phase. Current problems in engineering = Actual problems in machine building: 1 Materials Intern. scientific and practical. Conf., Novosibirsk, March 26, 2014 - Novosibirsk: Publishing House of the NSTU, 2014. - S. 482 - 488.

[17] EE Kornienko, Lapushkina E. Yu, V. Kuzmin et al. Air plasma spraying of powder coatings with gasdynamic focusing heterogeneous flow. Technology hardening, coating and repair: theory and practice: In 2 hours. Part 1 Materials 16th International Scientific and Practical Conference. - SPb .: Publishing House of the Polytechnic. Univ, 2014. - c. 78 - 86.

[18] VI Kuzmin, SN Grigoriev, OB Kovalev et al.'s New technology of plasma spraying of wear-resistant coatings. Friction and Wear, 2013 - T. 34. - № 3. - S. 221 - 226.

[19] V.I. Kuzmin, A.A. Mikhal'chenko, E.V. Kartaev, etc. The technique of formation of the axisymmetric heterogeneous flow for thermal spraying of powder materials. Journal of Thermal Spray Technology, 2011.

[20] Boronenko MP, Gulyaev IP, Gulyaev PY et al. Methods of controlling the temperature and velocity of the particles of the condensed phase in the plasma-arc spraying. Basic research, 2013. - № 10-6. - C. 1194 - 1199.

[21] GOST 23.208-79. Ensuring durability of products. Method of test materials to wear by friction of the abrasive particles loosely attached. 1979.

Journal of Physics: Conference Series **567** (2014) 012011 doi:10.1088/1742-6596/567/1/012011

Acoustic waves in polydispersed bubbly liquids

D A Gubaidullin, D D Gubaidullina, Yu V Fedorov

Institute of Mechanics and Engineering, Kazan Science Center, Russian Academy of Sciences, 2/31 Lobachevsky str., Kazan 420111, Russia

Kazan (Volga Region) Federal University, 18 Kremlyovskaya str., Kazan 420008, Russia

E-mail: kopperfildd@yandex.ru

Abstract. The propagation of acoustic waves in polydispersed mixtures of liquid with two sorts of gas bubbles each of which has its own bubble size distribution function is studied. The system of the differential equations of the perturbed motion of a mixture is presented, the dispersion relation is obtained. Equilibrium speed of sound, low-frequency and high-frequency asymptotes of the attenuation coefficient are found. Comparison of the developed theory with known experimental data is presented.

1. Introduction

Various problems of acoustics in mixtures of liquids with bubbles of gas or vapor are considered in monographs [1], [2]. Fundamental physical processes involved in bubble dynamics are described in books [3], [4]. Some aspects of propagation of acoustic waves in bubbly liquids are presented in the book [5]. Commander and Prosperetti [6] reviewed the formulation of a rigorous model for the propagation of pressure waves in bubbly liquids. The theoretical results for the phase speed, attenuation and transmission coefficient through a layer of bubbly liquid were compared with available experimental data. The propagation of acoustic waves in a mixture of a liquid with monodispersed bubbles [7] and in a mixture of a liquid with two sorts monodispersed bubbles [8] is investigated. In this work the dynamics of weak perturbations in mixtures of liquid with two groups of polydispersed gas bubbles of different sizes and various compositions is considered. The developed theory is compared with known experimental data.

2. The theoretical model

2.1. Basic equations

The set of linearized differential equations determining the propagation of sound waves in a mixture of liquid with two groups of polydispersed gas bubbles has the form:

$$\frac{\partial \rho_1'}{\partial t} + \rho_{10}\frac{\partial v'}{\partial x} = 0,\ \frac{\partial \rho_{2a}'}{\partial t} + \rho_{20}^a\frac{\partial v'}{\partial x} = 0,\ \frac{\partial \rho_{2b}'}{\partial t} + \rho_{20}^b\frac{\partial v'}{\partial x} = 0$$

$$\frac{\partial n_a'}{\partial t} + n_0^a\frac{\partial v'}{\partial x} = 0,\ \frac{\partial n_b'}{\partial t} + n_0^b\frac{\partial v'}{\partial x} = 0,\ \rho_{10}\frac{\partial v'}{\partial t} + \frac{\partial p_1'}{\partial x} = 0$$

$$\left\langle \frac{\partial p_{2a}'}{\partial t} \right\rangle_a = -3\gamma_{2a}p_0\left\langle \frac{1}{a}\frac{\partial a'}{\partial t} \right\rangle_a + \langle q_a p_{2a}' \rangle_a$$

$$\left\langle \frac{\partial p'_{2b}}{\partial t} \right\rangle_b = -3\gamma_{2b} p_0 \left\langle \frac{1}{b}\frac{\partial b'}{\partial t} \right\rangle_b + \langle q_b p'_{2b} \rangle_b \tag{1}$$

$$\frac{\partial a'}{\partial t} = w'_{Aa} + w'_{Ra}, \quad \frac{\partial b'}{\partial t} = w'_{Ab} + w'_{Rb}$$

$$a\frac{\partial w'_{Ra}}{\partial t} + 4\nu_1 \frac{w'_{Ra}}{a} = \frac{p'_{2a} - p'_1}{\rho^\circ_{10}}, \quad b\frac{\partial w'_{Rb}}{\partial t} + 4\nu_1 \frac{w'_{Rb}}{b} = \frac{p'_{2b} - p'_1}{\rho^\circ_{10}}$$

$$w'_{Aa} = \frac{p'_{2a} - p'_1}{\rho^\circ_{10} C_1 (\alpha^a_{20})^\beta}, \quad w'_{Ab} = \frac{p'_{2b} - p'_1}{\rho^\circ_{10} C_1 (\alpha^b_{20})^\beta}, \quad p'_1 = C_1^2 \rho_1'^\circ$$

$$\langle h \rangle_l = \frac{1}{\rho^l_{20}} \int\limits_{l_{\min}}^{l_{\max}} N_0(l) g_0(l) h dl = \int\limits_{l_{\min}}^{l_{\max}} N_0(l) g_0(l) h dl \Big/ \int\limits_{l_{\min}}^{l_{\max}} N_0(l) g_0(l) dl$$

$$g_0(l) = \frac{4}{3}\pi l^3 \rho^\circ_{2l}, \quad q_l = 3(\gamma_{2l} - 1)(i\omega)\frac{(y_l \coth y_l - 1)}{y_l^2}, \quad y_l = \sqrt{-\frac{i\omega l^2}{\kappa_{2l}}}, \quad l = a, b$$

Hereafter, the subscripts 1 and 2 refer to the parameters of the liquid and gas phases, respectively. The primes designate the perturbations of parameters, and 0 means the initial unperturbed state. The variables with the subscript a refers to gas bubbles of radius a, those with the subscript b refers to gas bubbles with a different radius b, x is the coordinate, t is the time, ρ° and ρ are the true and average density of the mixture, v is the velocity, p is the pressure, n is the number of bubbles per unit volume, γ is the adiabatic exponent, w is the bubble radial-motion velocity, α is the volume content, ν_1 is the kinematic viscosity of the liquid, ω is the frequency of perturbations, κ is the thermal diffusivity, C_1 is the speed of sound in the carrying phase, $N_0(a)$ and $N_0(b)$ are the gas-bubble-size distribution functions. In Ref.[9], the value of $\beta = 1/3$; as is shown below, in some cases it is better to use $\beta = 1/6$.

We investigate the solutions of this set of equations having the form of progressive waves for perturbations ϕ' ($\phi' = \rho'_1, \rho'_{2a}, \rho'_{2b}, p'_1, T'_1 ...$):

$$\phi' = A_\phi \exp[i(K_* x - \omega t)], K_* = K + iK_{**} \tag{2}$$

$$C_p = \omega / K, i^2 = -1$$

where K_* is the complex wave number, K_{**} is the attenuation coefficient, and C_p is the phase velocity.

2.2. Dispersion relation

From the condition of existence of the nontrivial solution of type (2) for set (1) of linear equations, the following dispersion relation is obtained:

$$\left(\frac{K_*}{\omega}\right)^2 = \frac{1}{C_f^2} + \frac{3\alpha^a_{20}\alpha_{10}\rho^\circ_{10}\langle Q_a \rangle_a}{3\gamma_{2a}p_0 - \langle Q_a S_a \rangle_a} + \frac{3\alpha^b_{20}\alpha_{10}\rho^\circ_{10}\langle Q_b \rangle_b}{3\gamma_{2b}p_0 - \langle Q_b S_b \rangle_b} \tag{3}$$

$$C_f = \frac{C_1}{\alpha_{10}}, Q_l = 1 + \frac{q_l}{i\omega}, S_l = \frac{i\omega l^2 h_l \rho^\circ_{10}}{1 + h_l t_l}$$

$$h_l = \frac{4\nu_1}{l^2} - i\omega, t_l = \frac{l}{C_1 (\alpha^l_{20})^\beta}, l = a, b$$

The low-frequency and high-frequency asymptotes of the attenuation coefficient K_{**} valid for the frequencies $\omega \ll \kappa_{2l}/l_{5,3}^2$ and $\omega \gg \kappa_{2l}/l_{4,3}^2$, respectively, are deduced from dispersion relation (3), where

$$l_{i,j} = \left[\frac{\langle l^i\rangle}{\langle l^j\rangle}\right]^{1/(i-j)}, i \neq j, l = a, b$$

Here $l_{i,j}$ are the average bubble radii [10], [11]. When obtaining the asymptotes, we neglected the kinematic viscosity ν_1 of the liquid.

The low-frequency asymptote is written as follows:

$$K_{**}^{(0)}(\omega) = \frac{C_e}{2}\left(\frac{\alpha_{10}\alpha_{20}^a(\gamma_{2a}-1)\rho_{10}^{\circ}a_{5,3}^2}{15p_0\gamma_{2a}\kappa_{2a}} + \frac{\alpha_{10}\alpha_{20}^b(\gamma_{2b}-1)\rho_{10}^{\circ}b_{5,3}^2}{15p_0\gamma_{2b}\kappa_{2b}}\right)\omega^2$$

$$C_e = \left(\frac{1}{C_f^2} + \frac{\rho_{10}^{\circ}\alpha_{10}\alpha_{20}^a}{p_0} + \frac{\rho_{10}^{\circ}\alpha_{10}\alpha_{20}^b}{p_0}\right)^{-1/2}$$

where C_e is the equilibrium speed of sound. It can be seen that these are the average radii $a_{5,3}$ and $b_{5,3}$ previously introduced during the analysis in Ref.[10, 11] of the acoustics of gas suspensions that are characteristic in this asymptote and describe the interphase heat exchange. In the propagation of sound waves, the attenuation coefficient is directly proportional to the square of the frequency of perturbations.

The high-frequency asymptote has the following form:

$$K_{**}^{(\infty)}(\omega) = C_f\frac{(H_1+H_2)}{2} - C_f^3(H_1+H_2)\frac{H_3}{4\omega^2}$$

$$H_1 = \frac{3\alpha_{20}^a\alpha_{10}h_a}{a_{4,3}}, H_2 = \frac{3\alpha_{20}^b\alpha_{10}h_b}{b_{4,3}}, h_a = \left(C_1(\alpha_{20}^a)^{\beta}\right)^{-1}$$

$$h_b = \left(C_1(\alpha_{20}^b)^{\beta}\right)^{-1}, H_3 = \frac{(H_1+H_2)^2C_f^2}{4} - (H_4+H_5)$$

$$H_4 = 3\alpha_{20}^a\alpha_{10}(\rho_{10}^{\circ} - 3p_0h_a^2\gamma_{2a})/(\rho_{10}^{\circ}a_{4,3}^2)$$

$$H_5 = 3\alpha_{20}^b\alpha_{10}(\rho_{10}^{\circ} - 3p_0h_b^2\gamma_{2b})/(\rho_{10}^{\circ}b_{4,3}^2)$$

3. Comparison between theory and experiment

The data shown in Figs. 1 and 2 enable us to compare the theory with the experimental data [12] for the dependences of the sonic speed and the attenuation coefficient on the perturbation frequency $f = \omega/(2\pi)$. Experimental data 1 were obtained for the values of $a_0 = 1.9\times10^{-3}$ m, $\alpha_{20}^a = 5.8\times10^{-4}$; theoretical curve 2 was obtained for ($a \in [1.9\times10^{-3}, 2\times10^{-3}]$, Fig.1) and ($a \in [1.9\times10^{-3}, 3\times10^{-3}]$, Fig.2), $\alpha_{20}^a = 5.8\times10^{-4}$, $\alpha_{20}^b = 0$, and disregarding the acoustic depression ($w'_{Aa} = 0$); and curve 3 was found at the same values as curve 2, but with taking into account the acoustic depression. The bubble-size distribution function is $N_0(a) = a^{-3}$. It can be seen that the acoustic depression substantially affects the behavior of the curves, which, in turn, agree well with the experimental data.

In real experiments, it is reasonably difficult to obtain a monodisperse bubble medium, when all bubbles have the same radius. It is very likely to have a group of very small bubbles with a low volume content, which renders, as is shown below, a substantial effect on the dynamics of propagation of waves.

In Fig.3a, we show experimental data 1 obtained for the following values: $a_0 = 2.6\times10^{-3}$ m and $\alpha_{20}^a = 0.01$; curve 2 corresponds to $a_0 = 2.6\times10^{-3}$ m and $\alpha_{20}^a = 0.01$; curve 3 corresponds

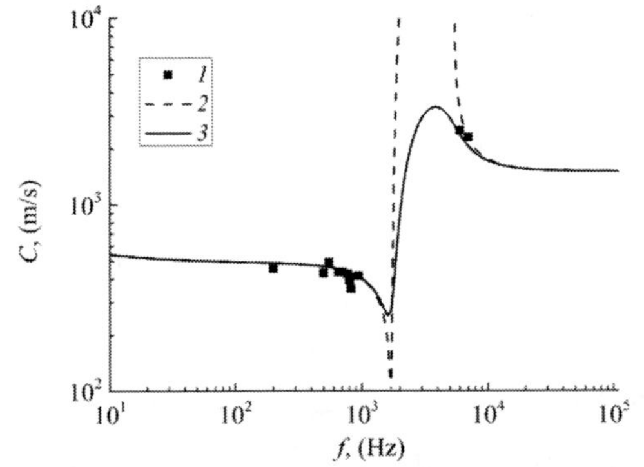

Figure 1. Comparison of the theory with the experimental data for the dependences of the sonic speed on the frequency of perturbations with taking into account the acoustic depression (solid line) and disregarding the acoustic depression (dashed line).

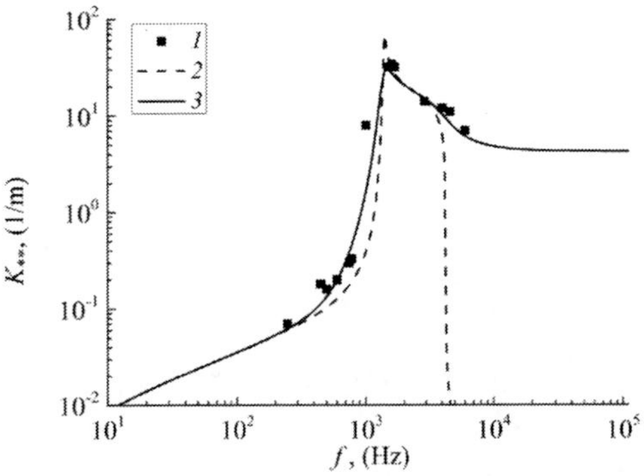

Figure 2. Comparison of the theory with the experimental data for the dependences of the attenuation coefficient on the frequency of perturbations with taking into account the acoustic depression (solid line) and disregarding the acoustic depression (dashed line).

to $\alpha_{20}^a = 0.01$, $a \in [2.6 \times 10^{-3}, 5.6 \times 10^{-3}]$, $\alpha_{20}^b = 2 \times 10^{-4}$ and $b \in [6.5 \times 10^{-4}, 8.5 \times 10^{-4}]$; and curve 4 corresponds to the same values as curve 3, but with the modified acoustic depression, which is presented below.

Experimental data 1 in Fig.3b are obtained for the following values: $a_0 = 2.1 \times 10^{-3}$ m, $\alpha_{20}^a = 0.0053$; curve 2 for $a_0 = 2.1 \times 10^{-3}$ m, $\alpha_{20}^a = 0.0053$; curve 3 for $\alpha_{20}^a = 0.0053$, $a \in [2.1 \times 10^{-3}, 4.1 \times 10^{-3}]$, $\alpha_{20}^b = 10^{-4}$, and $b \in [6.5 \times 10^{-4}, 8.5 \times 10^{-4}]$ m; curve 4 for the same values as curve 3, but with the modified acoustic depression.

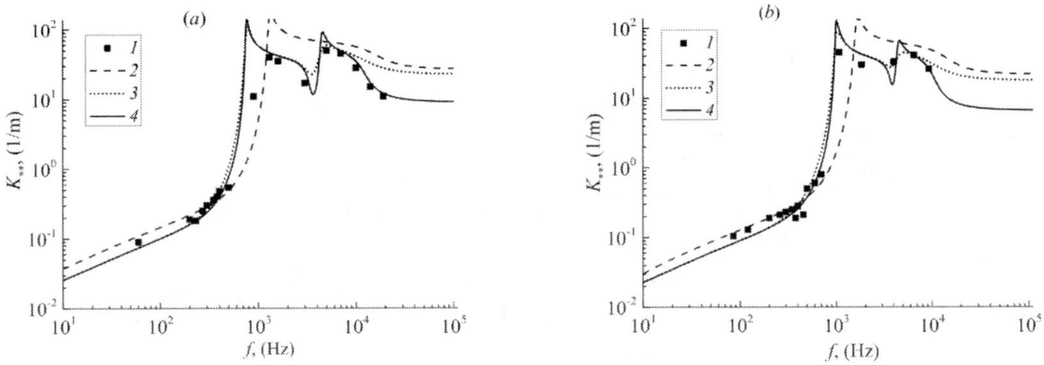

Figure 3. Comparison of the theory with the experimental data for the dependences of the attenuation coefficient on the frequency of perturbations.

The bubble-size distribution functions are

$$N_0(a) = a^2 \exp(-K_a a), K_a = 3/a_{5,3}$$

$$N_0(b) = b^2 \exp(-K_b b), K_b = 3/b_{5,3}$$

It is shown that the two-faction composition of the polydispersed bubble liquid results in the occurence of two peaks for the attenuation coefficient, which is observed in the experiment (Figs.3, curves 3 and 4). The experiment data show that the modification of the formula for the acoustic depressions w'_{Aa} and w'_{Ab} from set (1) by replacing the exponents $\beta = 1/3$ for $\beta = 1/6$ of the volume bubble contents α_{20}^a and α_{20}^b, respectively, results in an improvement of theoretical and experimental values of the attenuation coefficient (Fig.3, curves 4). The terms w'_{Aa} and w'_{Ab} are also determined from the solution of the problem on the acoustic depression of the spherical bubble in the carrying liquid [9].

4. Conclusions

Theory of the propagation of acoustic waves in polydispersed bubbly liquids is developed. It is shown, that the account of small bubbles with small volume content leads to a better agreement of theory and experiment.

Acknowledgment

This work was funded by the subsidy of the Russian Government to support the Program of Competitive Growth of Kazan Federal University among Worlds Leading Academic Centers.

References

[1] Nigmatulin R I 1991 *Dynamics of Multiphase Media* Vol. 1 (New York: Hemisphere) p 507
[2] Nakoryakov V E, Pokusaev B G and Shreiber I R 1993 *Wave Propagation in Gas - Liquid Media* (CRC Press; 2 edition) p 240
[3] Brennen C E 1995 *Cavitation and Bubble Dynamics* (Oxford: Oxford University Press) p 300
[4] Brennen C E 2005 *Fundamentals of Multiphase Flows* (Cambridge: Cambridge University Press) p 386
[5] Temkin S 2005 *Suspension Acoustics: An Introduction to the Physics of Suspensions* (Cambridge: Cambridge University Press) p 418
[6] Commander K W and Prosperetti A 1989 Linear pressure waves in bubbly liquids: Comparison between theory and experiments *J. Acoust. Soc. Am.* **85** 732–746
[7] Gubaidullin D A and Nikiforov A A 2010 Acoustic disturbances in a mixture of liquid with vapor-gas bubbles *High Temperature* **48** 170–175
[8] Gubaidullin D A, Nikiforov A A and Gafiyatov R N 2012 Acoustic waves in two-fraction bubble liquid with phase transformations *High Temperature* **50** 250–254
[9] Nigmatulin R I, Shagapov V Sh and Vakhitova N K 1989 Influence of Compressibility of Carrying Phase on Wave Propagation in Bubbly Liquids *Doklady Physics* **304** 1077–1081
[10] Gubaidullin D A and Nigmatulin R I 2000 On the theory of acoustic waves in polydispersed gas-vapor-droplet suspensions *Int. J. Multiphase Flow* **26** 207–228
[11] Gumerov N A and Ivandaev A I 1988 Sound propagation in polydispersed gas suspensions *J. Appl. Mech. and Tech. Phys.* **29** 706–714
[12] Silberman E 1957 Sound velocity and attenuation in bubbly mixtures measured in standing wave tubes *J. Acoust. Soc. Am.* **29** 925–933

About the basic laws of non-isothermal cross-flow past of a circular cylinder with beads mix

V L Fedyaev[1,2], I V Morenko[1] and E R Galimov[2]

[1] Institute of Mechanics and Engineering, Kazan Science Center, Russian Academy of Sciences, Russia
[2] Kazan National Research Technical University named after A N Tupolev, Russia

E-mail: morenko@imm.knc.ru

Abstract. The influence of mechanical impurities in the fluid on the structure of the flow near a stationary, rotating cylinder, on its hydrodynamic characteristics and the heat exchange with the stream was investigated numerically. It was shown that the presence of impurities in the liquid leads to a significant change in the pattern of the flow, to increase in the resistance and heat transfer of a stationary cylinder, reducing the lifting force. The emissivity of the rotating cylinder in a fluid flow with an admixture of substantially depends on the dimensionless velocity of the surface. If the speed is less critical, equal to 3.0, the heat is intensified. At higher values, on the contrary, the heat transfer deteriorates. The necessity of considering these circumstances when developing technologies for deposition of functional coatings is evidential.

1. Introduction

For many natural phenomena, technology is characterized by the flow of bodies of various shapes, including mobile, liquid containing impurities. Often the mechanical interaction of bodies with the environment is accompanied by heat transfer. The study of patterns of occurrence of these processes is of great interest.

An overview of the research methods of these processes, the obtained results are available in the literature [1-3]. Below, in relation to the technology of deposition of functional coatings on the surface, we give the results of numerical solution of non-isothermal cross-flow past a stationary, rotating circular cylinder monodispersed mixture. To simulate the motion of the medium is used as the Lagrangian and Euler continuum approach, the flow of viscous incompressible fluid is considered to be laminar. The complete system of equations with appropriate initial and boundary conditions is integrated using the finite volume method.

2. Governing equations

In the Cartesian rectangular coordinate system $x_1 0 x_2$ plane for carrier environment, representing a viscous incompressible fluid is described by the equations:

$$\frac{\partial u_i}{\partial x_i} = 0, \qquad (1)$$

$$\frac{\partial u_i}{\partial t}+u_j\frac{\partial u_i}{\partial x_j}=-\frac{1}{\rho}\frac{\partial p}{\partial x_i}+\nu\frac{\partial^2 u_i}{\partial x_j \partial x_j}-\frac{\overline{f}_{pi}}{\rho},\ i=1,2. \tag{2}$$

Here t is the time; u_i, u_j are components of the velocity vector $\boldsymbol{u}$ $(i,j=1,2)$; p is the pressure, ρ is the density of the fluid, ν is kinematic viscosity coefficient of the fluid, f_{pi} is the projection on the axis x_i averaged over volume forces hydrodynamic resistance of the particles.

The equation of conservation of energy is:

$$\frac{\partial}{\partial t}(\rho E)+\nabla\left(\vec{u}(\rho E+p)\right)=\nabla(\lambda\nabla T)-\overline{Q}_p, \tag{3}$$

where $E=h-\frac{p}{\rho}+\frac{u^2}{2}$, h is the enthalpy, T is the fluid temperature, λ is the thermal conductivity of the environment, $\overline{Q}_p$ is averaged by the amount of heat flux to the particles of impurities.

When considering the motion of particles of impurities, it is assumed that their concentration in the raw stream is small, the mutual influence of particle rotation is not taken into account. It is believed that particles are not deformed, have a spherical shape with the same radius r_p and average physical and mechanical properties, the material density of the particles ρ_p.

In the local Cartesian coordinate system $x0y$, the trajectory of the particle is described by the equations

$$\frac{dx}{d\tau}=u_{px},\ \frac{dy}{d\tau}=u_{py}, \tag{4}$$

where τ is the time; u_{px}, u_{py} are the projections of the velocity vector of a particle $\boldsymbol{u}_p$ at the coordinate axes. Given the hydrodynamic resistance force acting from the side of the carrier medium on the particle, gravity and Archimedes the equation of motion is represented in the form

$$\frac{d\boldsymbol{u}_p}{d\tau}=F_p\left(\boldsymbol{u}-\boldsymbol{u}_p\right)+\boldsymbol{g}\left(1-\frac{\rho}{\rho_p}\right). \tag{5}$$

Here $\boldsymbol{g}$ is the vector of gravity acceleration, $F_p=9C_p\,\mathrm{Re}_p\,\mu/12\rho_p d_p^2$, $d_p=2r_p$, $\mathrm{Re}_p=\frac{\rho d_p\left|\boldsymbol{u}_p-\boldsymbol{u}\right|}{\mu}$ is the Reynolds number of the particle, $C_p=a_1+\frac{a_2}{\mathrm{Re}_p}+\frac{a_3}{\mathrm{Re}_p^2}$ is the coefficient of hydrodynamic resistance, a_1, a_2, a_3 are constants.

According to relation (5) the power of the hydrodynamic resistance of the particles $f_p=0.5\pi C_p r_p^2\rho\left|\boldsymbol{u}-\boldsymbol{u}_p\right|\left(\boldsymbol{u}-\boldsymbol{u}_p\right)$.

Here we find averaged over the volume of the components of the resistance forces $\overline{f}_{pi}$ $(i=1,2)$, in the equation (2).

Since the sizes of the particles are small, they are long enough in the fluid flow, heat exchange with the environment is weak, it is assumed that the thermal state of particles is characterized by the average in terms of temperature T_p.

Provided that $\alpha_p=\overline{\alpha}_p=\mathrm{const}$, $T=\overline{T}=\mathrm{const}$, $\overline{T}$ is some time average temperature of the carrier medium

$$T_p=\overline{T}+\left(T_p^0-\overline{T}\right)\exp\left(-3\overline{a}_p\overline{B}\tau/\tau_0\right). \tag{6}$$

Equation (6) allows us to estimate the temperature change particles during their movement in the environment, as well as the heat exchange with the liquid medium characterized by specific surface heat flux

$$q_p = \left(\bar{B}\lambda_p / r_p\right)\left(T - T_p\right).$$

Hence, the heat flux to the particle of the impurities $Q_p = 4\pi r_p^2 q_p$. Making this value by the volume average, we get the value $\overline{Q}_p$ in equation (3).

If the impurity particles are thermally thin body, radiant heat transfer is not taken into account, the equation of thermal balance of the particles has the form

$$m_p c_p \frac{dT_p}{d\tau} = \alpha_p A_p \left(T - T_p\right),$$

where m_p is the mass, c_p is specific heat of the material, A_p is the particle surface area.

Recorded thus equations (1) - (3) must be supplemented by the boundary and initial conditions, and (4), (5) - initial. Let's discuss them below in relation to the specific geometry of the computational domain.

Let us choose the computational domain in the form of a rectangle the length of 0.10 m and a width of 0.04 m. We place the coordinate system origin $x_1 0 x_2$ into the center of the streamlined cylinder with radius 0.001 m in the destruction of 0.02 m from the input section, at an equal distance from the lateral boundaries.

In the input section perpendicular to the axis $0x_1$, we accept: $u_1 = u_\infty = \text{const}$, $u_2 = 0$; operating pressure p_0=0.1 MPa; the temperature of the carrier medium $T = T_0 = \text{const}$.

At the outlet of the computational domain, will use a "soft" boundary conditions: $\frac{\partial u_1}{\partial x_1} = 0$, $\frac{\partial u_2}{\partial x_1} = 0$, $\frac{\partial p}{\partial x_1} = 0$, $\frac{\partial T}{\partial x_1} = 0$, that means the alignment of both the hydrodynamic and thermal characteristics of the flow of the carrier medium.

On the lateral boundaries we place: $u_1 = 0$, $\frac{\partial u_2}{\partial x_2} = 0$, $\frac{\partial p}{\partial x_2} = 0$, $\frac{\partial T}{\partial x_2} = 0$.

It is believed that on the surface of the streamlined cylinder are implemented in terms of adhesion of the carrier medium $\boldsymbol{u} = 0$, if the body is motionless, and $\boldsymbol{u} = \boldsymbol{w}$, if the body rotates. The wall temperature remains constant, equal T_w. While specific heat flow is $q = \alpha\left(T_w - T\right)$.

In the initial moment of time $t = 0$ the environment instantly begins to move, $u_1 = u_\infty$, $u_2 = 0$, $p = p_0$, $T = T_0$.

When calculating the kinematics of particle impurities beginning of the local coordinate system $x0y$ is placed at the point of touch of the particles in the carrier medium. The starting point of these is located along the input section, the axis $0x$ is directed parallel to the axis $0x_1$. At the initial moment of time $\tau = 0$ $u_{px} = u_\infty$, $u_{py} = 0$, the temperature of the particles is $T_p = T_{p0} = T_0$. In addition to the speed of the particles in place of throw, the radius, the density of the particles, the mass flow rate G_p making influence on the number of particles coming in the calculated area per unit time are given.

By using Euler continuum approach given that mass exchange between the phases does not occur, the mass conservation equation can be written as:

$$\frac{\partial}{\partial \tau}\left(\varepsilon_i \rho_i\right) + \boldsymbol{v}_i \cdot \nabla\left(\varepsilon_i \rho_i\right) = 0, \quad i = 1,2; \tag{7}$$

here τ is the time, $\nabla = \boldsymbol{i}\frac{\partial}{\partial x_1} + \boldsymbol{j}\frac{\partial}{\partial x_2}$, $\boldsymbol{v}_i = v_{i1}\boldsymbol{i} + v_{i2}\boldsymbol{j}$ is the velocity vector i - phase; ε_1, ε_2 are volume concentration of the carrier and the dispersed phase ($\varepsilon_1 + \varepsilon_2 = 1$);

$$\frac{\partial}{\partial \tau}(\varepsilon_2) + \boldsymbol{v}_2 \cdot \nabla(\varepsilon_2) = 0. \quad (8)$$

The momentum equation for i - phase in the absence of gravitational forces has the form

$$\frac{\partial}{\partial \tau}(\varepsilon_i \rho_i \boldsymbol{v}_i) + \boldsymbol{v}_i \cdot \nabla(\varepsilon_i \rho_i \boldsymbol{v}_i) = -\varepsilon_i \nabla p_i + \nabla \cdot \Pi_i + \boldsymbol{R}_i. \quad (9)$$

Where p_i is the average pressure of the i - phase; $\Pi_i = \varepsilon_i \mu_i \left(\nabla \boldsymbol{v}_i + (\nabla \boldsymbol{v}_i)^T\right) + \varepsilon_i \left(\sigma_i - \frac{2}{3}\mu_i\right) \nabla \boldsymbol{v}_i \boldsymbol{I}$ is the stress tensor in the i - phase; μ_i, σ_i are shear, bulk viscosity; I is the unit tensor; $\boldsymbol{R}_i = \frac{(-1)^i}{V}\int_V \gamma(\boldsymbol{v}_2 - \boldsymbol{v}_1)dV$ is the averaged interaction force between phases, V is the control volume,

$\gamma = \varepsilon_2 \rho_2 f/\delta$, $\delta = 2\rho_2 a^2/9\mu_1$, $f = \frac{c_D \mathrm{Re}_{21}}{24}$,

$$c_D = \begin{cases} 24\left(1 + 0.15\mathrm{Re}_{21}^{0.687}\right)/\mathrm{Re}_{21}, & \mathrm{Re}_{21} \le 1000 \\ 0.44, & \mathrm{Re}_{21} > 1000 \end{cases},$$

$\mathrm{Re}_{21} = \frac{2\rho_1 |\boldsymbol{v}_2 - \boldsymbol{v}_1| a}{\mu_1}$ is the Reynolds number of the relative motion phases.

Equation of heat transfer in the carrier and dispersed phases

$$\frac{\partial}{\partial \tau}(\rho_i c_{pi} T_i) + \boldsymbol{v}_i \cdot \nabla(\rho_i c_{pi} T_i) = \nabla \cdot (\lambda_i \nabla T_i) + Q_i, \quad (10)$$

here $Q_i = \frac{(-1)^i}{V}\int_V 4\pi a^2 \varepsilon_2 \beta_{21}(T_2 - T_1)dV$, $\beta_{21} = \lambda_1 \mathrm{Nu}_{21}/2a$ is the heat transfer coefficient between the carrier and the disperse phases, the Nusselt number is calculated by the formula Ranz-Marshall: $\mathrm{Nu}_{21} = 2 + 0.6\mathrm{Re}_{21}^{1/2}\mathrm{Pr}_{21}^{1/3}$ $(0 \le \mathrm{Re}_{21} \le 200, 0 \le \mathrm{Pr}_{21} \le 250)$.

Basic equations (7) - (10) are supplemented by the boundary and initial conditions. The input section perpendicular to the axis $0x_1$, set: $v_{11} = v_{21} = w$, $v_{12} = v_{22} = 0$; operating pressure p_0=0.1 MПa; temperature of the carrier and the dispersed phase is the same: $T_1 = T_2 = T_0 = \mathrm{const}$; concentration $\varepsilon_2 = \varepsilon_0 = \mathrm{const}$.

At the outlet of the computational domain we use the following boundary conditions: $\frac{\partial v_{i1}}{\partial x_1} = \frac{\partial v_{i2}}{\partial x_1} = \frac{\partial p_i}{\partial x_1} = \frac{\partial T_i}{\partial x_1} = 0$ $(i = 1,2)$, $\frac{\partial \varepsilon_2}{\partial x_1} = 0$.

On the lateral boundaries: $\frac{\partial v_{i1}}{\partial x_2} = \frac{\partial p_i}{\partial x_2} = \frac{\partial T_i}{\partial x_2} = 0$, $v_{i2} = 0$ $(i = 1,2)$, $\frac{\partial \varepsilon_2}{\partial x_2} = 0$.

It is believed that on the surface of the cylinder attachment conditions implemented by the carrier and dispersed phases $\boldsymbol{v}_1 = \boldsymbol{v}_2 = 0$; there is a perfect thermal contact: $T_1 = T_2 = T_w$. In this heat flux $q_i = -\lambda_i \partial T_i/\partial n = \alpha_i\left(T_w - T_i^*\right)$ $(i = 1,2)$, here n – outward normal to the surface of the cylinder, T_i^* – temperature i-phase away from the cylinder (average temperature), α_i – required local convective heat transfer coefficient.

At the initial time $\tau = 0$ the fluid instantly begins to move, $v_{i1} = w$, $v_{i2} = 0$, $p = p_0$, $T_i = T_0$, $\varepsilon_2 = \varepsilon_0$.

3. Results

It is established that in the case of a stationary cylinder at the picture wrapping his dispersed mixture have a significant impact width of the jet; the concentration, size, material density of the particles (droplets) of the impurities (figure 1).

Comparing the data shown in figure 1 *a, b*, convinced that if the stream of particles impurities narrow, the particles are localized outside the cylinder in distinct vortex clusters. At the same mass concentration of particles γ_∞, the Stokes number $\mathrm{Stk} = u_\infty \rho_p d_p^2 / 18\mu D$ (u_∞ – flow velocity, D – the cylinder diameter; ρ_p, d_p are the density, the particle diameter of the impurities), when the size of the streamlined body is less than the width of the jet, such localization is not observed.

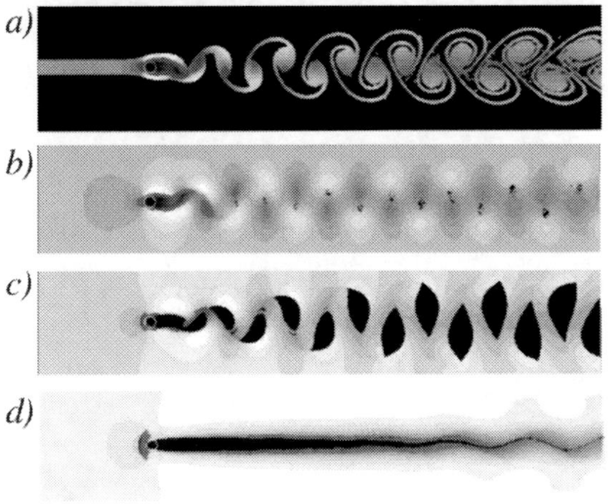

Figure 1. Module of the particle velocity of the impurities *a*) Stk=0.1, $\gamma_\infty = 23\%$; *b*) Stk=0.1, $\gamma_\infty = 23\%$; *c*) Stk=1.0, $\gamma_\infty = 23\%$; *d*) Stk=2.0, $\gamma_\infty = 60\%$.

If the width of the jet, the diameter of the cylinder, the mass concentration of particles are unchanged (figure 1 *b, c*), the increasing of the number of Stokes (density, particle size) leads again to their localization in the body, but in some different form (figure 1 *c*). Finally, the increase both the concentration and the number of Stokes radically changes the picture (figure 1 *d*). Moreover, at small Stokes numbers after the contact with the surface of the cylinder of solid particles roll down the windward side downstream, towards the vortex stagnant zone; and a high bounce off the surface, in the region behind the cylinder gets there is an extremely amount of them (figure 2).

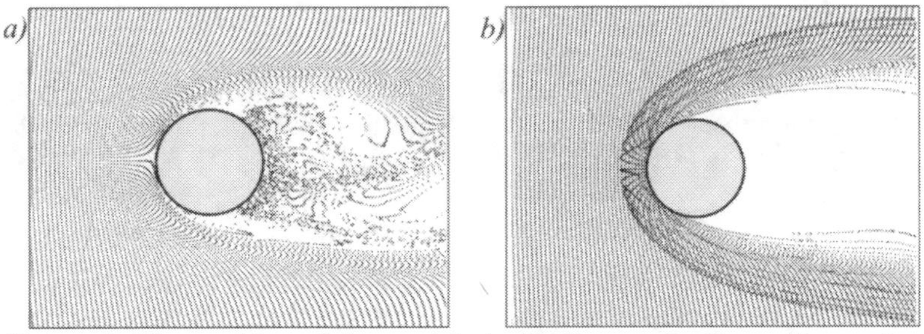

Figure 2. The position of the particle in the flow at the Reynolds number Re=100, $\beta = \rho_p / \rho = 4.6 \cdot 10^5$, γ_∞=0.23: Stk=0.1 (a); Stk=1.0 (b)

Journal of Physics: Conference Series **567** (2014) 012012 doi:10.1088/1742-6596/567/1/012012

At constant other parameters, depending on the relative density β of the volumetric, the concentration of particles α_2 on the surface of the cylinder varies considerably (figure 3). It is seen that when $\beta > 1$ (figure 3, curve 1) the concentration of the particles of impurities takes the maximum value in the frontal point of the cylinder, then the flow direction is reduced at the point of separation of the flow reaches its minimum. On the surface of the aft portion of the cylinder, the impurity concentration of particles is high enough, varies along the surface slightly. Thus, the surface area of the cylinder $\bar{L} = 0 \div \pi/2$ can be divided in three distinctive areas: frontal and stern area where a high concentration of impurities, and an intermediate region with a low concentration of impurity particles. At low relative density of the material particles of impurities $\beta < 1$ (figure 3, curve 2) their concentration along the surface remains almost unchanged.

It follows that to obtain the deposition of a uniform coating on the surface of a stationary circular cylinder we should strive to ensure that the material density of the particles (droplets) would be less than the density of the carrier medium.

It should be noted that in the case of the rotating cylinder instantaneous distribution of impurity concentration on the surface is approximately the same as on the surface of the stationary cylinder.

However, due to the rotation of the cylinder, the thickness of the coating layer, a lot depends on the nature of the distribution of particles on the surface.

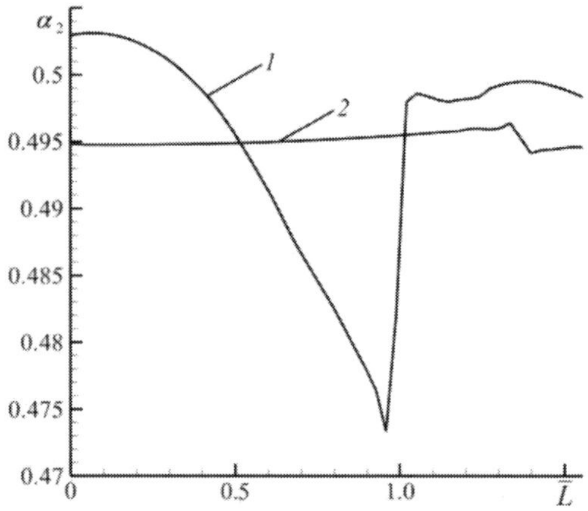

Figure 3. Changes in the concentration of the impurity particles along the surface of the cylinder ($\bar{L} = l/D$): 1 – β=21.5, 2 – β=0.

In addition to the thickness of the coating an important indicator of its quality is the continuity of the material and adhesion, characterized by the bond strength between the coating layer and the substrate. Both figures largely depend on the thermal processes occurring in the contact area of the rolling flow of the mixture and the treated body.

It is known that the addition of the gas solids improves the heat transfer of streamline bodies due to the decrease of the viscous sublayer. Indeed, if a circular cylinder is stationary, the local Nusselt number presented with increasing mass flow rate of particles of impurities increases (figure 4). In this case, as expected, the greatest heat transfer occurs from the windward side of the cylinder, in the vicinity of the separation point of the flow it is at a minimum, in the aft portion of the cylinder it increases.

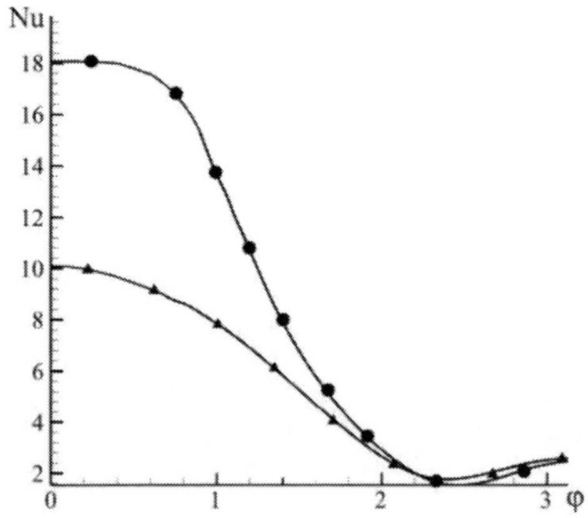

Figure 4. Local Nusselt number at Re=100, Pr=0.75, Stk=0.6, $\beta=4.6\cdot10^5$: mass flow rate of particles kg/s (●), kg/s (▲).

The thermal interaction between heterogeneous environment with a streamlined body on the heat exchange rate, defined medium on the surface of the Nusselt number $\overline{\mathrm{Nu}}$, is affected by the fact that near the surface concentration of impurities can increase substantially, which will change the conductivity of the mixture, the nature of the flow. From the calculations it follows that when the Reynolds number Re=100

$$\overline{\mathrm{Nu}}=\overline{\mathrm{Nu}}_0+1.7\cdot10^6\frac{\gamma_\infty(1-0.3/\mathrm{Stk})}{\mathrm{Stk}(\beta+(1-\beta)\gamma_\infty)},$$

where $\overline{\mathrm{Nu}}_0$ is the average Nusselt number around the cylinder in a uniform stream.

It was also established that at a fixed Reynolds number, the change in specific heat of the material of the particles does not have significant effect on heat transfer of a cylinder. However, the thermal conductivity of the material particles of the heat exchange cylinder with the flow of the mixture varies significantly, and the stronger, the larger the Reynolds number (figure 5).

The influence of particle size on heat transfer streamlined cylinder with the environment at a fixed particle concentration impurity at the entrance to the settlement area is also largely depends on the density and thermal conductivity of the material particles. If the density of the material particles is less than or close to the density of the carrier medium, and the thermal conductivity is greater, then on the condition that the particle diameter increases, the heat transfer deteriorates.

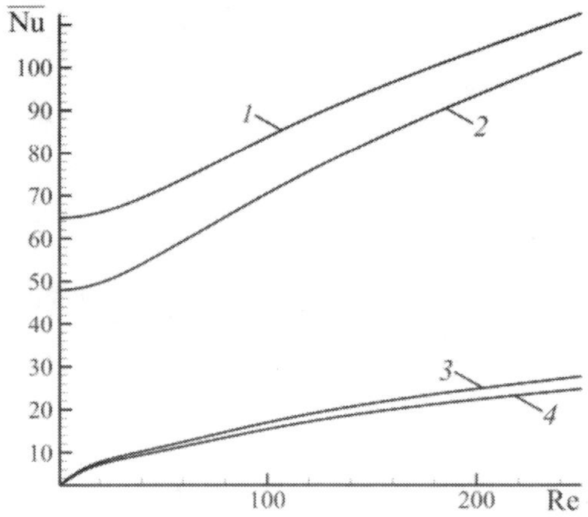

Figure 5. The dependence of the average Nusselt number from Reynolds number of the cylinder for different materials of the particles: 1 – gold, 2 – aluminum; 3 – water without particles; 4 – polypropylene.

In the case of a rotating cylinder with respect to heat transfer with two-phase flow can distinguish three characteristic change interval dimensionless speed of the surface of the cylinder $\alpha=0.5\omega D/u_\infty$

Journal of Physics: Conference Series **567** (2014) 012012 doi:10.1088/1742-6596/567/1/012012

(ω is the angular speed of rotation of the cylinder): $0<\alpha<3.0$, $3.0<\alpha<4.3$, $4.3<\alpha<4.9$, (figure 6). When $\alpha<3.0$ the throw of an impurity in a homogeneous viscous fluid flow several intensifies the heat exchange, the Nusselt number $\overline{Nu}$ increases by approximately 15%. A further increase in the speed of rotation of the cylinder leads to the fact that when $3.0<\alpha<4.3$ the heat transfer degrades significantly $\overline{Nu}$ reduced by almost half. Subsequent increase α $(4.3<\alpha<4.9)$ decrease rate of heat transfer decreases, the Nusselt number to flow with the admixture approach Nusselt number is presented for homogeneous flow.

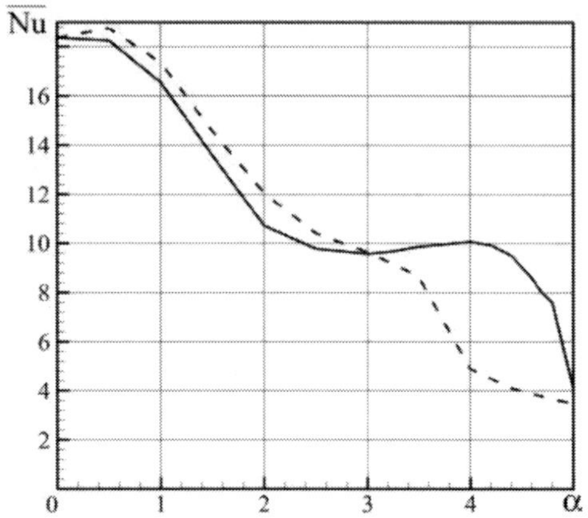

Figure 6. The dependence of the average Nusselt number of the cylinder against relative rotation speed: solid line – homogeneous flow, dashed line – the flow with particles.

The identified patterns are of great practical importance, because they allow to evaluate the impact on the quality of the coatings of the main indicators of the sprayed materials, enable rational selection of technological modes.

References

[1] Morenko I V and Fedyaev V L 2010 Features of two-phase flow around a circular cylinder *Ecological Bulletin of the scientific centers of the black sea economic cooperation* **4** 52–58

[2] Morenko I V and Fedyaev V L 2011 Nonisothermal flow around a circular cylinder of the monodispersed mixture *Thermal processes in engineering* **3**(6) 242–252

[3] Morenko I V and Fedyaev V L 2014 Laminar nonisothermal flow of a viscous fluid with solid particles past a rotating circular cylinder *Journal of Engineering Physics and Thermophysics* **87**(3) 566–572

The study the erosion of the electrodes under the influence moving electric arc

A T Gabdrakhmanov, I H Israphilov, A T Galiakbarov

Department of Energy and Information, Kazan Federal University, Naberezhnye Chelny, Russian Federation

E-mail:veyron000@mail.ru

Abstract. The paper presents the results of experimental study of the processes of interaction of a moving arc with electrodes pulsed plasma generator in the range of parameters I = 60÷500 A, U = 30÷70 V, G = 0÷155 l/min, d_E = 4÷14 mm and L_E = 2÷20 mm. In this article the results of the study traces left by moving arc on the electrode surfaces are described, and the impact velocity of the arc on the form and sizes these traces, ranging V_D = 7÷130 m / s. We show the research results of the heating and erosion of the electrodes under the influence of high-current arc moving.

1. Introduction

Near-electrode effects in plasmatron with moving arc a significant impact on the formation and burning of electric arc in a pulsed plasma generator.

Near-electrode effects - the processes in gas discharges in the plasma with inhomogeneous concentration, temperature, and other parameters, concluded between the electrode and the almost homogeneous plasma. In contrast to the homogeneous plasma positive column where the current flows under the influence of an electric field in the near-electrode parts importanceprocesses the transfer of charged particles due to diffusion and under the influence of a temperature gradient. Very close to the electrode distribution the electron and the ion by speed are generally different from the Maxwell distribution.

The complexity of the near-electrode effects is determined not only the variety of conditions in which they take place, but also a necessity in many cases regarded as effects near the electrode, and in the electrode at the mutual influence on each other.This fact characterizes the composition and properties of the electrode plasma. For example, the existence and self-renewal of pulsed arc defined the formation electrode spots and of erosion material electrode. Consequently, one of the main proposition in the study of near-electrode effects, is research of the spots left by moving arc on the electrode surface [1, 3].

2. Experimental studies of spots

In the pulsed plasma generator the arc ignited at the end of the electrode, it is run by a magnetic field and leave traces on the surface of the electrodes.

Journal of Physics: Conference Series **567** (2014) 012013 doi:10.1088/1742-6596/567/1/012013

We give a qualitative description of the traces (Fig. 1). Depending on the geometric dimensions (*d*), the interelectrode gap (*L*) and the mode of arc traces at the electrodes are essentially different. For example, for high currents ($I \geq 150$A) and velocities ($V \geq 20$m / sec) (Figs. 1 b and c) traces on the cathode are a plurality of grooves parallel to the direction of movement of the arc. Obviously, while there are a large number of spots which have a high mobility and move with an arc. When moving many spots run in the footsteps of other spots.

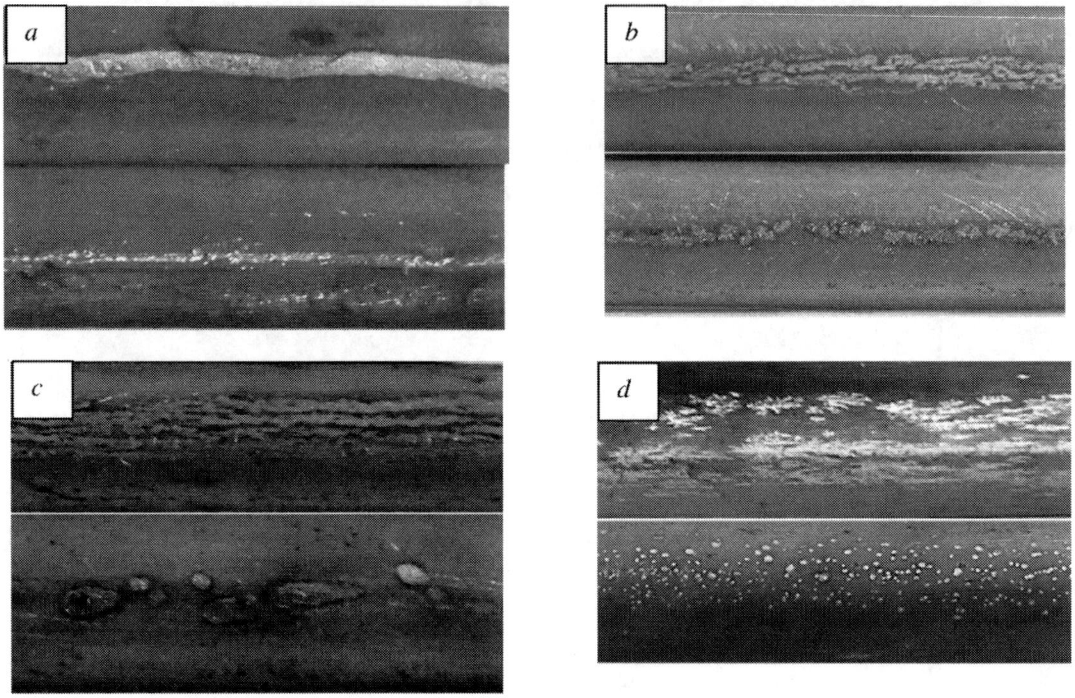

Figure 1. Photographs of traces on the electrodes from steel 20 (from above cathode): *a*- d=8mm, L=8mm, I=120A, U=46V, V=7,5m/s; *b* - d=8 mm, L=4 mm, I=230A, U=36V, V=25m/s; *c* - d=8 mm, L=4 mm, I=380A, U=38V, V=35m/s; *d* - d=8 mm, L=5 mm, I=1500A, U=70V, V=130m/s; U – arc voltage

Flow pattern of the electrode spots movement greatly influence the temperof the flow in the conducting channel of the arc.Trajectory of motion spots, probably approaching particle trajectories conducting gas in the arc column. Spots move in the direction of the arc, while run away laterally from the axis of symmetry and disappear, reaching the edge of the arc column. Besides the basic branching on the cathode surface seen small branching. The formation of this small structure, presumably due to the high mobility microspots. Tends to form microspots at the edges fused spots where due to the destruction of electrode material formed sharp edges and leads to increased gradients of the electric field, which facilitates electron emission.

At low speed of the arc up to 10 m/s traces are continuous (Fig. 1 a). By increasing the speed of movement on the surface of electrode, the arc leaves traces of irregular shape which do not have the preferred orientation (Fig. 1 b - d). Because of the oxide film near electrode spots have low mobility, moving behind the arc. As a result of the breakdown between the arc and the wall formed a new spot and between spots remain gaps. At low speeds, trace is small. With increasing speed discontinuities between separate traces increase and reach 5 mm. It is connected that at lower speed the air between the arc and the wall succeeds to warm more and shunt between the electrode and the wall of the arc occurs at a lower breakdown voltage, i.e. with a smaller displacement of the arc relative to a substantially stationary spot.

Journal of Physics: Conference Series **567** (2014) 012013 doi:10.1088/1742-6596/567/1/012013

At the anode, for small speeds, there is also the continuity of the trace (Fig. 1 a). By increasing the speed of the arc gaps appear between spots, which have a circular shape (Fig. 1 b and d) or a form of the drop, elongated in the direction of motion (Fig. 1 c). The number of spots increases with increasing current, and the size of each spot is greatly reduced.

On the copper electrode has a similar tendency, i.e. at low speeds of movement of the arc, traces are continuous, with increasing velocity traces consist of separate spots. With increasing speed traces consist of separate spots. But because of a lower melting point of copper erosion surface more intensively.

3. The erosion of the electrodes

The study the erosion of the electrodesto determine service lifeof the electrodes and the plasmatron as a whole. To determine the erosion of the electrodes is necessary to estimate the amount of heat introduced into the electrodes. To measure the heat flux supplied to the electrodes used in the K-type the thermocouple and the multimeter, thermocouples close up in the electrodes. The study of thermal fields in the electrodes was carried out using analytical methods [4].

As a result of running a single arc electrode temperature was increased byΔT. By raising the temperature determines the amount of heat given from the arc to the electrode. After reaching the maximum temperature of the thermocouple changed slightly for a few seconds, which suggests that there is no leakage of heat. In Figure 2 is a plot of heat from the arc at the electrode from current, which it is seen that the amount of heat introduced to the anode far exceeds the heat losses at the cathode. Consider the reasons for such behavior.

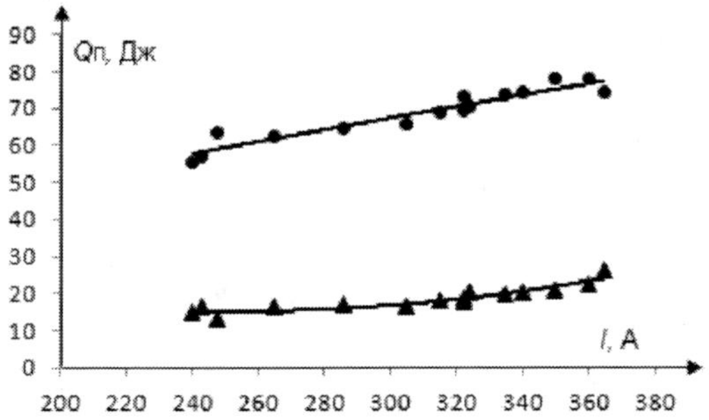

Figure 2.Dependence the heat to the electrodes ($Q_{П}$) of the arc current (I), the electrode material Steel 20d=8mm, L=8mm; •-anode, ▲-cathode

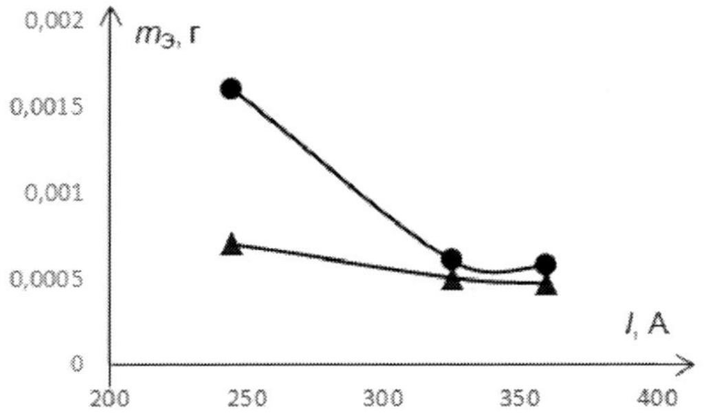

Figure 3.Dependence the erosion of the electrodes ($m_Э$) of the arc current (I), the electrode material Steel 20d=8mm, L=8mm; •-anode, ▲-cathode

The arc moving under its own magnetic field belong to the species of the arc with cold cathode and the cathode spot. The current in such arcs flows through one or many small, fast and disorderly moving appearing and disappearing spots on the cathode. The current density is very high in spots, $J_k = 10^4\text{-}10^7$ A/cm^2. For a short time the localization of spots metal in this location is strongly heated, broken, vaporized but next to the spots and generally remains relatively cold cathode.

In the spot enters the ion current. Ions bring kinetic energy, which acquired in the cathode fall. Together with the potential energy released during neutralization, it goes into heating the metal. This, together with the strong field causes the emission and on the other hand, the evaporation of the metal. Couples create in the case of a vacuum gap the environment which is ionized emitted electrons accelerated in the cathode fall. This, in turn, is the source, the ion current. The evaporation process of localized hot center depends on the outflow direction heat energy to adjacent layers of metal and other components of energy balance.

The voltage drop in the anode region is created as a result of the extraction of electrons from the arc column and acceleration when entering them into the anode. In the anode region has substantially only an electron current, due to the small amount of negatively charged ions having lower speed than the electron. An electron gets on surface of the anode, gives the metal not only kinetic energy but also the energy of the work function, so an anode receives energy from the arc column, not only in the form of electron beam, but in the form of thermal radiation. Consequently, the anode temperature is always higher, and it generates more heat.

Quantitative evaluation of erosion of the electrodes was carried out by determining the mass change of the electrodes under the influence of a moving arcing using laboratory scale (Fig. 3).

The analysis found that in said range with increasing current decreases erosion of the electrodes and thus increase the number of pulses which produce of pulsed electrode of the plasma generator prior to their destruction. The reason for this is that during the growth of the current speed increases, changing the impact of the arc on the electrodes and thereby reduces the erosive effect of moving the arc. Shared resource electrodes, for example, current 360A is 50 hours.

4. The conclusion

The arc in the electrode region exists due to thermionic emission of the atoms and electrons with the surface, which ultimately affects the stability of the arc. In turn, the severe erosion of the electrodes can reduce the life of the plasmatron. In this connection, it is necessary to select the operating mode of plasmatron such that movement of the arc accompanied by micro grafting. On the mode of movement of the arc strongly influenced the speed of the arc, thus changing the speed of the arc, you can choose the best mode of arc, get a steady stream of plasma on output and increase the service life of theplasmatron [5,6].

References

[1] Jasko O.I. *The electric arc in a plasmatron.* - Minsk: Science and Technology, 1977. 149s.

[2] Koroteev A.S., Mironov V.M., Svirchuk Y.S. *Plasma torches: design, characterization, calculation.* - M .: Mechanical Engineering, 1997.-296 p.

[3] Jasko O.I.*Interaction with the arc plasmatron electrodes.* - Minsk: Publishing "Science and Technology", 1982. 152 p.

[4] Israfilov I.H., Bashmakov D.A. Galiakbarov A.T., Mandrik P.A., Ganiev M.M.*Analytical description of the temperature field in the material under the influence of combined heat source* // News of higher educational institutions. Physics. Volume 57, № 3-3, 2014., p. 148-151.

[5] Galiakbarov A.T., Nugumanova A.I., Zvezdin V.V. Gabdrakhmanov A.T., Saubanov R.R.*Investigation of the influence of parameters of pulsed plasma generator on quality manufacturing process* // Bulletin KSTU. AN Tupolev. Number 2, 2010, s.50-52.

[6] Israfilov I.H., Saubanov R.R., Rakhimov R.R.*Prospective application of highly concentrated energy for surface heat treatment products* // Socio-economic and technical systems: research, design, optimization. Number 1, 2011, p. 25-30.

Research of Radio-Frequency Electric Discharge in Electrolyte in the Processes of Surface Treatment

Al F Gaisin[1], I Sh Abdullin[2], R Sh Basyrov [1], S Y Petriakov[1]

[1]Faculty of Physics and Mathematics, Kazan National Research Technical University named after A.N.Tupolev -KAI, Kazan, 420111, Russia
[2]Faculty of Nanotechnologies and Nanomaterils, Kazan National Research Technological University, Kazan, 420015, Russia

E-mail: almaz87@mail.ru

Abstract. The object of this research is radio-frequency electric discharge in electrolyte in the processes of surface treatment of materials. The forms and structures of electric discharge, the results of X-ray structure analysis, metallographic analysis of surface samples before and after treatment are presented in the article.

1. Introduction
At present time the electro physical processes in gas between solid electrodes are studied very well. The low temperature plasma of electric discharges finds wide applications in different branches of industry. One of the new ways of plasma production is radio-frequency current supply to electrodes immersed into the electrolyte [3]. Electric discharges in liquid electrolyte could be used for surface treatment of different materials by plasma technologies. Traditional surface treatment technologies could be supplemented with plasma-electrolyte surface treatment [1, 2].
A number of features of radio-frequency electric discharge in the electrolyte in the processes of surface treatment are presented in this article.

2. Experiment
The study was performed on the installation consisting of two ten litter volume electrolyte cells, high-frequency generator HFG8-60/13 which tuned to frequency 13.56 MHz and liquid electrolyte control system. The main parameters of experiment: the voltage is 3500-4500 V. The saturated solution of NaCl in tap water was used as electrolyte. The electrical steel used for production electrical machine magnetic circuits, apparatus and devices was chosen as a sample for surface treatment. The total mass fraction of alloying elements (silicon and aluminum) is from 2.8 to 3.8%. The pictures of discharges were shot by Panasonic DMC-FZ20 digital camera. X-ray diffraction analysis was carried out on a diffract meter SIEMENS D5000. The anode material of X-ray tube is Mo with wavelength K_α (λ= 0.7093 Å). The surface micro-relief was studied by Olympus BX51 microscope.

3. Results
Figure 1 shows the discharge development and it's interaction with sample surface. Steam gas shell with micro discharges cover and treat the sample surface. Cathode spots appear on the electrolyte

surface. The main contribution to surface treatment is due to anodic dissolution and ion impact. The processing time is 120 s.

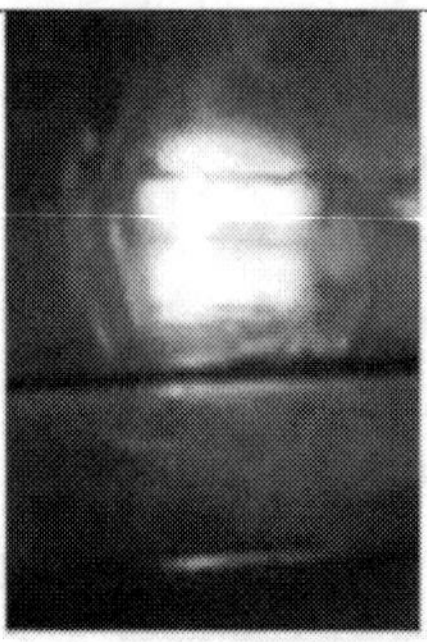

Figure 1a. The radio-frequency electric discharges in electrolyte in the processes of surface treatment. The voltage is 3500 V. Liquid electrolyte is a saturated solution of NaCl in tap water.

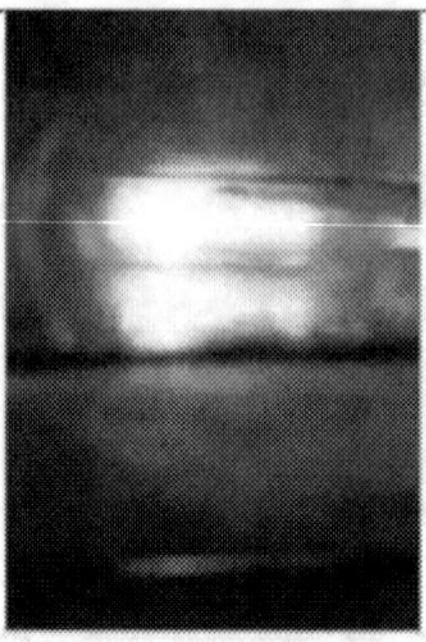

Figure 1b. The radio-frequency electric discharges in electrolyte in the processes of surface treatment. The voltage is 4000 V. Liquid electrolyte is a saturated solution of NaCl in tap water.

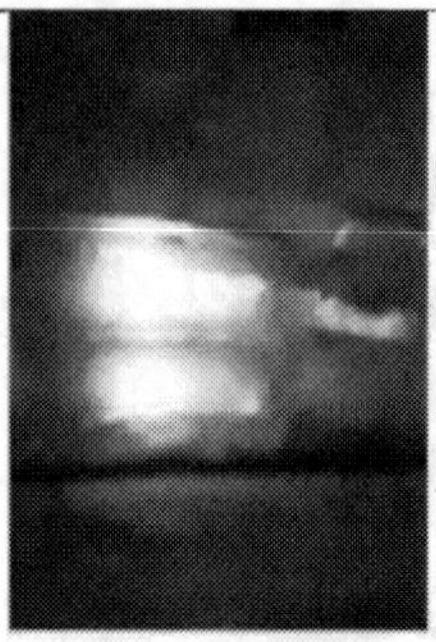

Figure 1c. The radio-frequency electric discharges in electrolyte in the processes of surface treatment. The voltage is 4500 V. Liquid electrolyte is a saturated solution of NaCl in tap water.

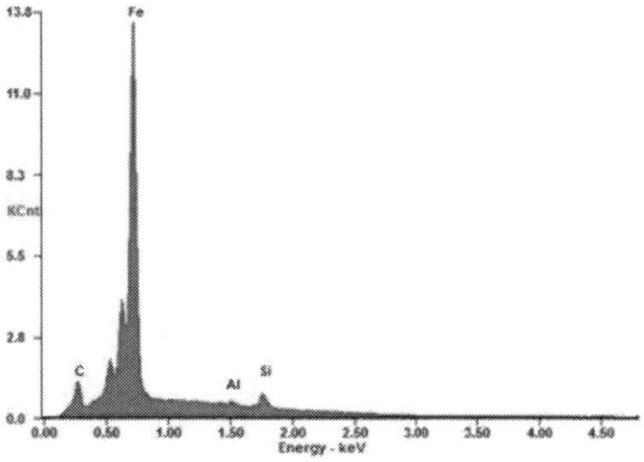

Figure 2a. X-ray spectrum (energy-dispersive) of virgin sample.

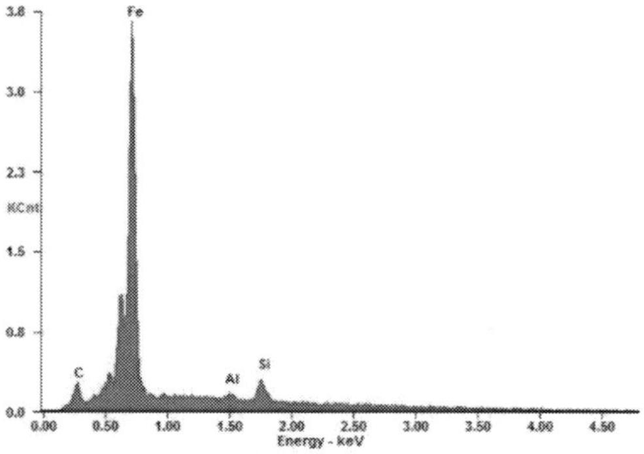

Figure 2b. X-ray spectrum (energy-dispersive) of the sample treated by electric discharge in electrolyte.

The X-ray spectrum analysis on Figure 2 shows that samples before and after treatment have the same chemical composition. The samples contain elements such as iron (Fe), carbon (C), silicon (Si) and aluminum (Al). Figure 3 shows that diffraction patterns of samples before and after treatment are very similar, this indicates the identity of their structures. Reference data was used as reference diffraction patterns. Comparing the X-ray diffraction spectra of samples (Figure 3) it can be said that the ferrite is in its master phase since the greatest reflexes belong to it, besides all spectra except №1 (initial sample) show the presence of ε-Fe3C phase (ε-iron carbide). On the spectra there are several unrecognized reflexes which due to low content of components are not sufficient for unambiguous diagnostics. Theoretical and experimental relative intensities differ greatly from each other since the examined sample was in the form of a solid matter and not in the form of a polycrystalline powder. It can be concluded that the intensity was strongly influenced by the texture of the sample as well as the static fluctuations caused by coarse crystallites, drift of device and counter, absorption factor.

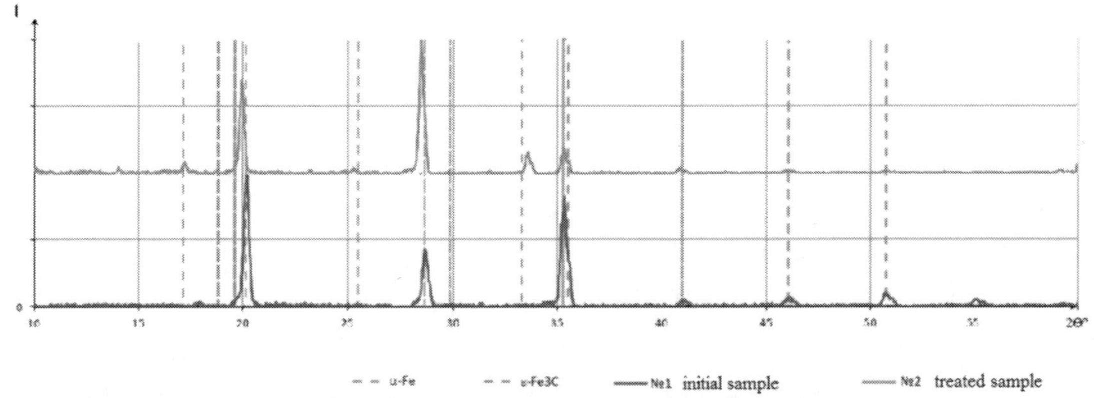

Figure 3. Comparison of experimental and reference diffraction patterns (α-Fe and ε – Fe3C).

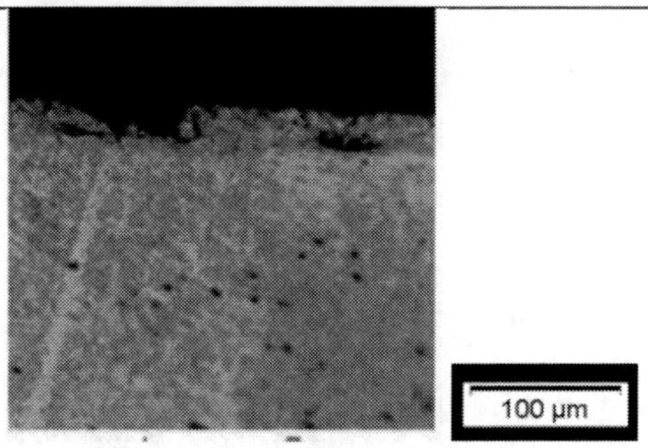

Figure 4a. Samples micro-relief before treatment.

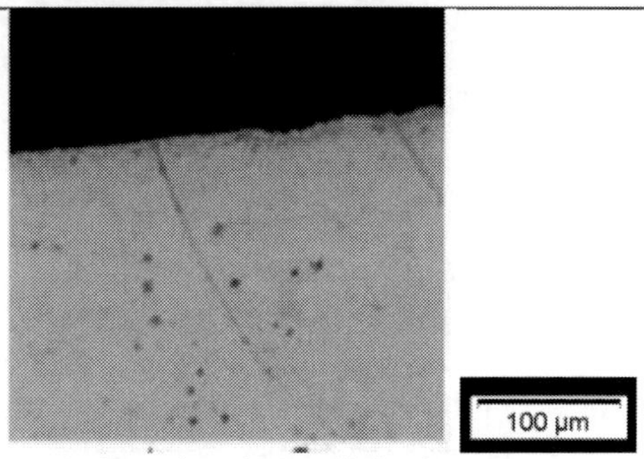

Figure 4. Sample micro-relief after treatment.

Figure 4 shows roughness and granularity decreasing of samples structure.

References

[1] Gaisin A.F. 2013 High Temperature **6** 863 – 866
[2] Gaisin A.F., Sarimov L.R. 2011 Plasma Physics Reports **6** 535 – 540
[3] Dautov G, Dautov I, Fayrushin I and Kashapov N 2013 J. Phys.: Conf. Series **479** 012001

LTP Coatings 2014 IOP Publishing
Journal of Physics: Conference Series **567** (2014) 012015 doi:10.1088/1742-6596/567/1/012015

A calculation of characteristics of creeping discharge using Tepler model for electrical conductivity

I. G. Galeev, B. A. Timerkaev, R. Z. Fayzullin
Kazan National Research Technical University named by A.N.Tupolev, Kazan, 420111 Russia Kazan, K.Marx str., 10

E-mail: galeev.physics@kstu-kai.ru

Abstract. The work is devoted to the modeling of discharge creeping over the dielectric surface. Non-stationary system of differential equations is used for describing the creeping discharge in various forms. There is presented a finite-difference approximation of differential equations and boundary conditions. The resulting system is solved numerically by the method of driving. There are received space distributions of such characteristics as the distribution of current in the interelectrode space, the charge passed through certain section at different points of time. The obtained curves are characteristic for uncompleted stage of discharge, and for the complete stage creeping discharge.

The study of the creeping discharge at the present time is an important task [1]. In this work is considered one-dimensional unsteady model of the creeping discharge. In a first approximation, it is based on the wave propagation of current and voltage along a line with distributed parameters - capacitance and resistance. In this case the change of the voltage U and current I at each point of the channel x is described by the equations

$$U_x = -\rho \cdot I, \qquad (1),$$

$$I_x = -c \cdot U_t, \qquad (2)$$

where c and ρ - capacitance, and the resistance per unit length of the discharge; $0 \leq x \leq l$.

The relationship between resistance and current can be represented in the form $\rho = \rho_0 \cdot \left(1 + \frac{\rho_0 \cdot Q}{k}\right)^{-1}$, where ρ_0 - resistance in front of the discharge, k = (1÷2)10^{-4} Ohm Cl/cm , $Q = \int_0^t I dt$. Level of ionization in creeping discharge is still small and the nature of the conductivity is determined by the collisions of electrons with neutral gas. Conductivity in this case itself almost depends only on a degree of ionization. It is assumed a applicability of the model of Tepler, which gives a good qualitative description of the resistance of the spark, while the energy released in the channel, still not great. Also it is supposed the increase of electrical conductivity proportional to the current.

From equations (1) and (2) it is possible to receive a equation of the second order with respect to Q

$$Q_{xx} = x\rho_0 Q_t \left(1 + \frac{\rho_0}{k} Q\right)^{-1}. \qquad (3)$$

Boundary conditions can be obtained by integrating equation (2)

$$Q_x(0,t) = -cU(0,t),\ Q_x(l,t) = 0.$$

In the unfinished stage of discharge voltage can be considered as a linear function of time

$$U(0,t) = U_m \frac{t}{t_m},\ 0 \le t \le t_m;\ U(0,t) = U_m,\ t \ge t_m.$$

Equation (3) with initial and boundary conditions can be represented in dimensionless form

$$q_{yy} = \frac{\beta q_\tau}{1+\beta q},\ q_y(0,\tau) = \begin{cases} -\frac{\tau}{\tau_m}, 0 \le \tau \le \tau_m, \\ -1,\ \tau \ge \tau_m, \end{cases} \quad q_y(1,\tau) = 0, \quad (4)$$

where $y = \frac{x}{l}$, $\tau = \frac{tU_m}{kl}$, $q = \frac{Q}{clU_m}$, $\beta = \frac{1}{k} U_m c\rho_0 l$ and accordingly the current in dimensionless units $i = \frac{Ik}{c\cdot(U_m)^2}$. In real conditions $\beta = 10 \div 10^3$, and already at very small $q \sim 1/\beta$ equation (4) and his decision almost cease to depend on β.

Equation (4) was solved by the method of finite-difference approximations. For this purpose the following notation could be introduced:

$$q_j = q(\tau, y), \qquad q_j^* = q(\tau + \Delta\tau, y)$$
$$q_{j+1} = q(\tau, y + \Delta y), \quad q_{j+1}^* = q(\tau + \Delta\tau, y + \Delta y)$$
$$q_{j-1} = q(\tau, y - \Delta y), \quad q_{j-1}^* = q(\tau + \Delta\tau, y - \Delta y)$$

Then the equation (4) takes the form:

$$q_{j+1}^* - \left(\frac{\Delta y^2}{\Delta\tau}\cdot\frac{1}{\left(\frac{1}{\beta}+q_j\right)} + 2\right)\cdot q_j^* + q_{j-1}^* = -\frac{\Delta y^2}{\Delta\tau}\cdot\frac{1}{\left(\frac{1}{\beta}+q_j\right)}\cdot q_j;$$

The boundary condition is written as $\frac{\partial q}{\partial y}\big|_{y=0} = f(\tau)$. Accordingly, in finite-difference representation, it is represented in the form $-3q_1^* + q_2^* = -(2q_1 - f_*\Delta y)$.

Typical results of numerical simulation are presented in Fig. 1-3.

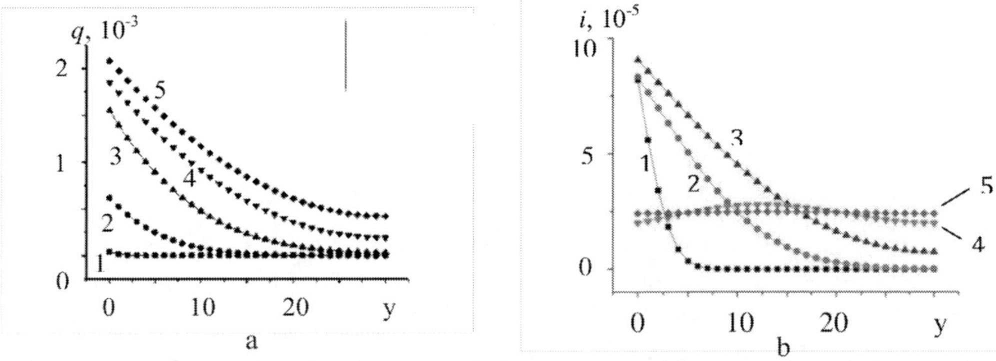

Fig. 1 Graphs of q (a) and i (b) in the discharge gap at different points of time τ when $\beta = 1.5*10^2$. The curves correspond to: 1 - τ =0,511; 2 - τ = 7.48; 3 - τ =18,4; 4 - τ =29,7; 5 - τ =39,8.

Figure 1. b represents the data of calculations with $\beta = 1.5*10^2$. It shows a passage of the wave current in the electrode gap. The electric current i, generated on the high -voltage electrode, extends over time to the other electrode. Accordingly, the values of charge q are increased (Fig. 1 .a). Figure 2. a, b represents data of calculations with $\beta = 1.5*10^3$. At some point of time the current on the initiating electrode is terminated, however, wave power even extends to the opposite electrode. Figure 3. a, b shows the distribution q and i with $\beta = 1.5*10^2$ along the discharge gap at different points of time. There is also observed the current wave displacement to the opposite electrode.

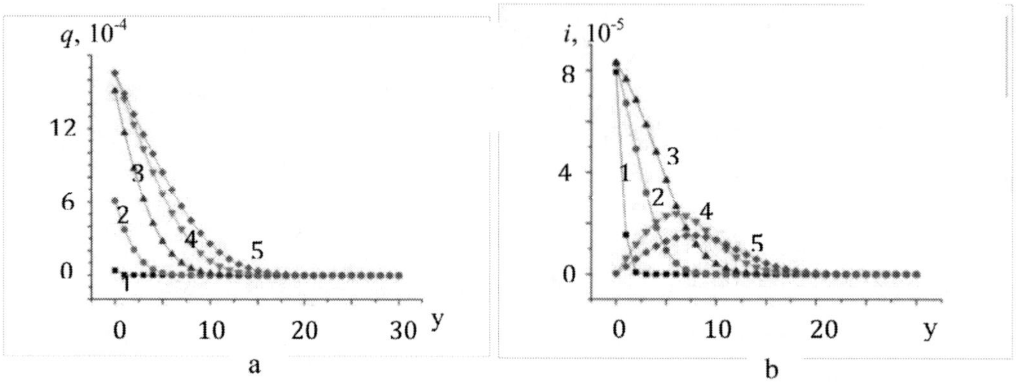

Fig. 2 Graphs of q (a) and i (b) in the discharge gap at different points of time T when β = $1.5*10^3$. The curves correspond to: 1 - τ =0,511; 2 - τ =of 7.48; 3 - τ =18,4; 4 - τ =29,7; 5 - τ =39,8.

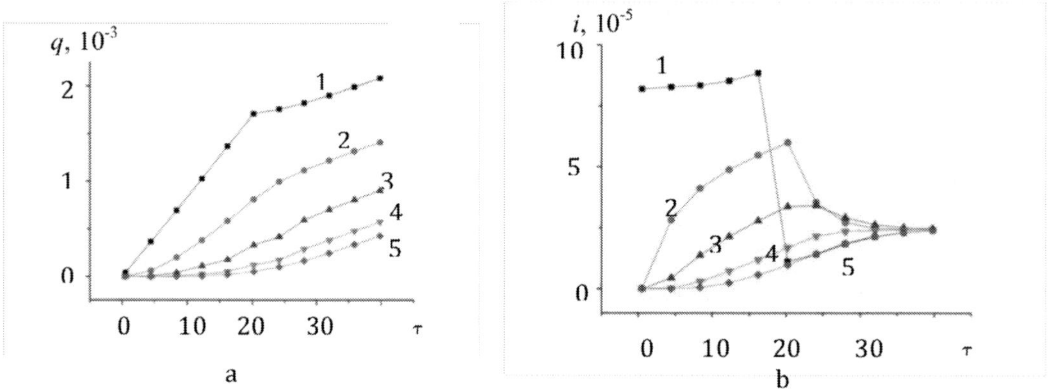

Fig. 3 Calculated dependence of q (a) and i (b) from time τ at different points of the discharge gap at β = $1.5*10^2$. The curves correspond to: 1 - y =1; 2 - y =8; 3 - y =15; 4 - y =22; 5 - y =29.

Retrieved characteristic temporal distribution of the current flowing through high-voltage initiating electrode, qualitatively consistent with the known experimental data.

References

[1] Mesjac G A 2004 Pulse technics and electronics M: Nauka 704 p.

Propagation of Acoustic Waves in Liquid Containing Multilayer Barrier

D A Gubaidullin[1,2] and A A Nikiforov[1,2]

[1]Institute of Mechanics and Engineering, Kazan Science Center, Russian Academy of Sciences, 2/31 Lobachevsky str., Kazan 420111, Russia
[2]Kazan (Volga Region) Federal University, 18 Kremlyovskaya str., Kazan 420008, Russia

E-mail: anikiforov1@yandex.ru

Abstract. The dynamics of acoustic waves in a liquid containing multilayer barrier is studied. Using the subroutine of the fast Fourier transform the calculation of the distortion of the acoustic signal at the diagnosis of a multilayer sample comprising a layer of a liquid with polydispersed bubbles is executed. A good agreement between theoretical and experimental data is obtained.

Introduction

It is well known that even an insignificant (on volume or mass) number of bubbles in a liquid strongly affects the acoustic properties of the liquid. The fundamentals of mechanics and thermal physics of bubble liquids and the most essential results on studying the wave processes in such media are presented in [1, 2]. The conditions under which the carrying-liquid compressibility manifests itself for the problems of acoustics of bubble liquids are established in [3]. The results of experimental investigations of bubble-gel samples are given in [4]. In [5, 6], the propagation of weak harmonic perturbations in two-fraction mixes of a liquid with polydisperse bubbles of gas of different composition was investigated. We study pressure-pulse propagation in a polydisperse bubble medium on the basis of [6]. The dependence of the complex wave number on the vibration frequency according to [6] provided that the disperse phase consists only of bubbles of one gas; i.e., the volume content of the second-fraction bubbles is zero and has the form

$$\left(\frac{K_*}{\omega}\right)^2 = \frac{1}{C_f^2} + \frac{3\alpha_{20}\alpha_{10}\rho_{10}\langle Q\rangle}{3\gamma_2 p_{10} - \langle QS\rangle} \qquad (1)$$

$$C_f = C_1/\alpha_{10},\ \ Q = 1 - 3(\gamma_2 - 1)(y\cdot\coth(y) - 1)/y^2,\ \ y = \sqrt{i\omega a^2/\kappa_2},$$

$$S = i\omega a^2 h_a \rho_{10}/(1 + h_a t_a),\ \ h_a = -i\omega + 4\nu_1/a^2,\ \ t_a = a/(C_1(\alpha_{20})^{\beta}),$$

$$\langle Q\rangle = \frac{1}{\rho_2}\int_{a_{min}}^{a_{max}} N(a)g_0(a)Q\,da,\ \ \langle QS\rangle = \frac{1}{\rho_2}\int_{a_{min}}^{a_{max}} N(a)g_0(a)QS\,da,$$

$$\rho_2 = \int_{a_{min}}^{a_{max}} N(a) g_0(a) da, \quad g_0(a) = \frac{4}{3}\pi a^3 \rho_{20},$$

where and p_{10} are the true density and pressure of unperturbed liquid, C_1 is the sound speed in liquid, ν_1 is the kinematic viscosity of liquid, ρ_{20} is the true density of gas in bubbles, γ_2 is the adiabatic exponent of gas in bubbles, $N(a)$ is the distribution function over bubble sizes a, κ_2 is the thermal diffusivity of gas, and c_{p2} is the specific heat capacity of gas in bubbles at constant pressure.

Results

Further, we calculated the dynamics of the pulse perturbation of the pressure in liquid containing the multilayer barrier and compared the results with the experiment. In [4], a sample consisting of two polycarbonate layers and the layer containing industrial gel with polydisperse bubbles was investigated experimentally.

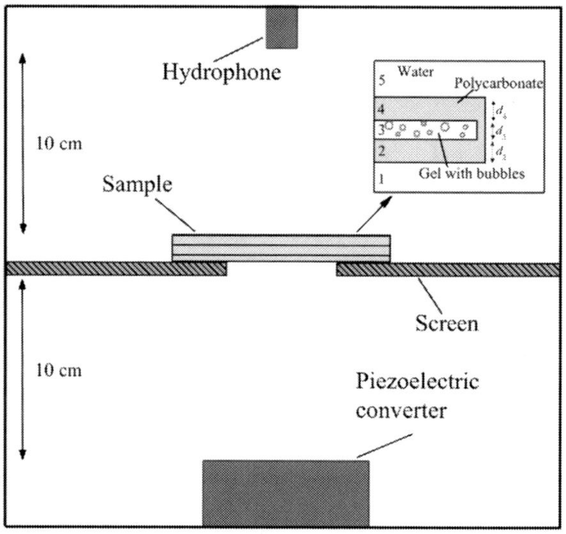

Figure 1. Scheme of installation for acoustic diagnostics from experimental data [4].

The scheme of the experiment is shown in figure 1. The sample is placed in a vessel with water between an acoustic-signal source (the piezoelectric converter) and a hydrophone. On the basis of the analysis of the parameters of sound waves transmitted through a sample, the data on its properties are found. The function of the distribution of bubbles over sizes was determined as

$$N(a) = \frac{n_t}{\sqrt{2\pi}\varepsilon} \exp\left\{-\frac{[\ln(a/a_0)]^2}{2\varepsilon^2}\right\}, \quad a_{min} \le a \le a_{max}.$$

Where $a_0 = 0.08\,\text{mm}$, $\varepsilon = 0.05$ is the distribution parameters [4]. For approximating the initial pulse from the experiment, the pressure pulse perturbation is set with the help of the imaginary part of the Morlet wavelet with the splash frequency $\nu_B = \omega_B / 2\pi = 0.25\,\text{MHz}$:

$$p'(0,t) = \text{Im}\left\{\exp[-i2\pi\nu_B(t-\tau)]\exp\left[\frac{-(t-\tau)^2}{2\delta^2}\right]\right\}. \quad (2)$$

The parameters $\tau = 20\,\mu\text{s}$ and $\delta = 3\ \mu\text{s}$ determine the splash location and width, respectively.

According to [7], the result of transmission of a plane monochromatic wave $\sim \exp[i(Kx - \omega t)]$ through a multilayer sample is a plane wave $\sim W\exp[i(Kx - \omega t)]$, where W is the wave-transmission coefficient, which is determined through the layer impedances Z_i and the input layer-boundary impedances Z_i^{in}:

$$W = \frac{Z_4^{in} + Z_4}{Z_4^{in} + Z_5} e^{iK_4 d_4} \frac{Z_3^{in} + Z_3}{Z_3^{in} + Z_4} e^{iK_3 d_3} \frac{Z_2^{in} + Z_2}{Z_2^{in} + Z_3} e^{iK_2 d_2} \frac{2Z_1}{Z_1 + Z_2} . \quad (3)$$

$$Z_j = \rho_j \frac{\omega}{K_j}, \quad Z_j^{in} = \rho_j \frac{Z_{j-1}^{in} - iZ_j \operatorname{tg}(K_j d_j)}{Z_j - iZ_{j-1}^{in} \operatorname{tg}(K_j d_j)} Z_j, \quad Z_1^{in} = Z_1 .$$

Here, d_j and K_j are, respectively, the thickness and wavenumber of the *j*-th layer (for the liquid layer with bubble inclusions $d_3 = h$, $K_3 = K_*$).

Using the subroutine of the fast Fourier transform (FFT), we calculate the evolution of pressure pulse (2). With the help of the inverse Fourier transform, we determined the complex amplitudes of the harmonic components of initial pressure pulse (2) at $x = 0$. Then, using equation (3), we determine the amplitudes of harmonic components of the pulse transmitted through a multilayer sample to the hydrophone (at $x = \sum d_j$). The variation of pressure as a function of time in a new position is calculated with the help of the direct Fourier transform.

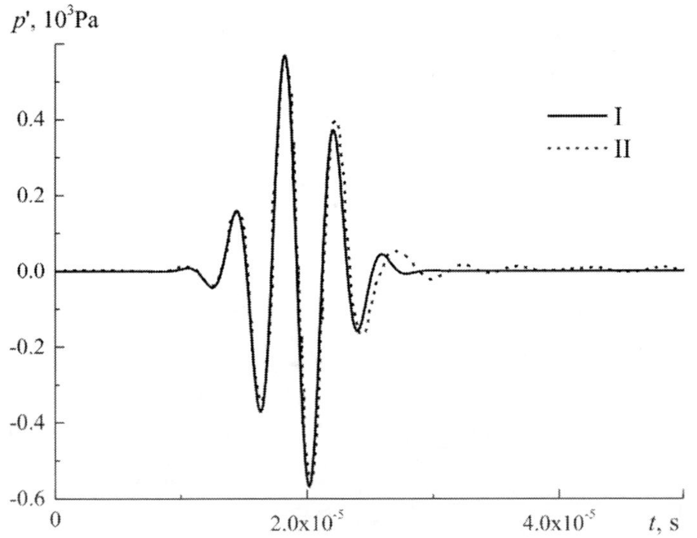

Figure 2. Pulse perturbation of pressure in purified water (*I* is the calculation, and *II* is the experimental data [4]).

The results of calculation of evolution of the pressure pulse perturbation were compared with the experimental results [4]. The data in figure 2 for the initial signal show that function (2) describes well the initial acoustic signal written in the experiment [4]. The number of harmonics in the subroutine FFT was selected from the condition of the pulse motion in purified water (when there is no dispersion or dissipation) without distortion.

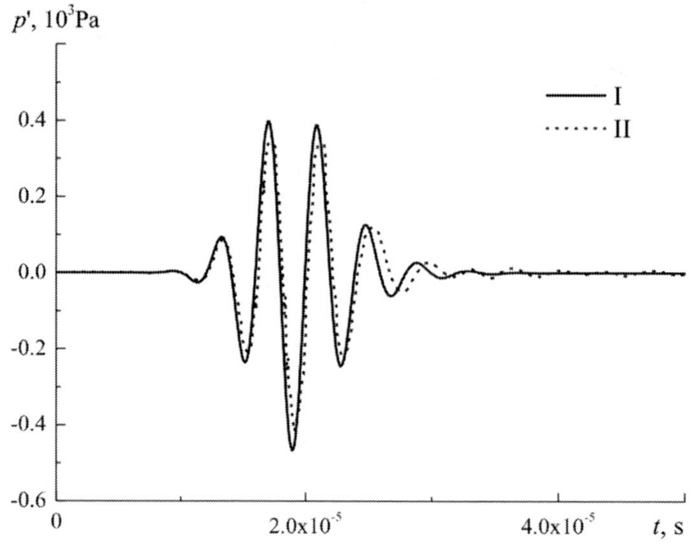

Figure 3. Pulse perturbation of pressure in water containing a multilayered sample (*I* is the calculation, and *II* is the experimental data [4]).

The experimental results [4] shown in figure 3 are obtained at the following bubble-layer parameters: $h = 2\,\text{mm}$, $\alpha_{20} = 0.0015$, $a_0 = 0.08\text{mm}$, and correspond to the average distance $l = a_0 / \alpha_{20}^{1/3} = 0.7\,\text{mm}$ between bubbles. In this case, the wavelength under consideration on a pure liquid is $L = 2\pi C_1 / \omega = 15\,\text{mm}$. The continuum model under consideration for a bubble liquid is valid for $L >> l$ and $h >> l$. Although the second condition is not completely fulfilled, the theoretical curve of the dependence of the pressure at the hydrophone on the reduced time agrees well with the experiment.

Thus, the theory under consideration can be used confidently for calculation of the distortion of the acoustic signal in the diagnostics of multilayer samples with bubble inclusions.

Acknowledgments
This work was funded by the subsidy of the Russian Government to support the Program of Competitive Growth of Kazan Federal University among World's Leading Academic Centers.

References

[1] Nigmatulin R I 1991 *Dynamics of Multiphase Media* (New York: Hemisphere)
[2] Nakoryakov V E, Pokusaev B G and Shreiber I R 1993 *Wave Propagation in Gas-Liquid Media* (New York: CRC Press)
[3] Nigmatulin R I, Shagapov V Sh and Vakhitova N K 1989 *USSR Academy of Sciences Doklady* **304** 1077
[4] Leroy V, Strybulevych A, Page J H and Scanlon M G 2008 *J. Acoust. Soc. Am.* **123** 1931
[5] Nigmatulin R I, Gubaidullin D A, Fedorov Yu V 2012 *Doklady Physics* **57** 451
[6] Nigmatulin R I, Gubaidullin D A, Fedorov Yu V 2013 *Doklady Physics* **58** 261
[7] Brekhovskikh L M 1980 *Waves in layered media* (New York: Academic Press)

Numerical simulation of Schlichting streaming induced by standing wave in rectangular enclosure

D A Gubaidullin[1,2], P P Osipov[1,2] and R R Nasyrov[1,2]
[1] Institute of Mechanics and Engineering, Russian Academy of Sciences, 2/31, Lobachevsky str, 420111, Kazan, Russia.
[2] Kazan (Volga region) Federal University, 18, Kremlin str, 420008, Kazan, Russia.

E-mail: petro300@rambler.ru

Abstract. Acoustic streaming of incompressible fluid induced by standing wave in rectangular computational region is simulated. The impact of width of computational region on Schlichting vortices is analyzed. The Crank-Nickolson scheme across region is used to solve Prandtl equations. The upper boundary of computational region simulates the extern flow via Neumann homogeneous condition. The mean velocity field over large number of cycles is computed. It is shown that to achieve proper mean velocity and appearance of Schlichting vortices the width of computation region must be more than 8 times of thickness of acoustic boundary layer.

Numerical simulation of acoustic streaming generated by resonant oscillations in an enclosure is an interesting topic due to various applications of wave separation technologies. The mathematical model of acoustic streaming and related references can be found in [1, 2, 3].

1. Governing equations

For fluid velocity Prandtl equations are used

$$\frac{\partial u}{\partial t}+u\frac{\partial u}{\partial x}+v\frac{\partial u}{\partial y}-\nu\frac{\partial^2 u}{\partial y^2}=\frac{\partial U}{\partial t}+U\frac{\partial U}{\partial x} ,$$

$$\frac{\partial u}{\partial x}+\frac{\partial v}{\partial y}=0 ,$$

where $U=U(x,t)$ is fluid velocity far from boundary $y=0$, ν is kinematic viscosity of fluid. Further the standing wave

$$U=U_0\sin k_w x\sin\omega t , \qquad \frac{\partial U}{\partial t}=\omega U_0\sin k_w x\cos\omega t , \qquad \frac{\partial U}{\partial x}=k_w U_0\cos k_w x\sin\omega t ,$$

in rectangular region $0\le x\le L$, $0\le y\le H$ is considered, where U_0 is amplitude of wave, $k_w=\dfrac{2\pi}{\lambda}$ is wave number, $\lambda=2L$ is wave length, $T=\dfrac{2\pi}{\omega}$ is period.

Boundary conditions.

Left boundary $(0 \le y \le H)$: $u(0,y,t) = U(0,t) = 0$, $v(0,y,t) = 0$.

Right boundary $(0 \le y \le H)$: $u(L,y,t) = U(L,t) = 0$, $v(L,y,t) = 0$.

Upper boundary: $\left.\dfrac{\partial u}{\partial y}\right|_{y=H} = 0$.

Lower boundary: $u(x,0,t) = 0$, $v(x,0,t) = 0$.

Find mean velocity of acoustic streaming (l is enough large integer)

$$\bar{u} = \frac{1}{lT}\int_0^{lT} u(x,y,t)\,dt, \qquad \bar{v} = \frac{1}{lT}\int_0^{lT} v(x,y,t)\,dt .$$

2. Numerical scheme

After introducing a regular grid $x_i = i\Delta x$, $(i = 0,\ldots,n)$, $\Delta x = L/n$,

$y_j = j\Delta y$, $(j = 0,\ldots,m)$, $\Delta y = H/m$,

the Crank-Nicolson scheme across the boundary layer can be written

$$\frac{u_{i\,j}^{k+1} - u_{i\,j}^{k}}{\Delta t} - \gamma \frac{\left(u_{i\,j-1}^{k} - 2u_{i\,j}^{k} + u_{i\,j+1}^{k}\right) + \left(u_{i\,j-1}^{k+1} - 2u_{i\,j}^{k+1} + u_{i\,j+1}^{k+1}\right)}{2\Delta y^2} = f_{i\,j}^{k+1/2}$$

where $f_{i\,j}^{k+1/2} = \left\{\dfrac{\partial U}{\partial t} + U\dfrac{\partial U}{\partial x}\right\}^{k+1/2} + g_{i\,j}^{k+1/2}$. In this paper a first-order approximation is used

$$g_{i\,j}^{k+1/2} = -u_{i\,j}^{k}\frac{u_{i+1\,j}^{k} - u_{i-1\,j}^{k}}{2\Delta x} - v_{i\,j}^{k}\frac{u_{i\,j+1}^{k} - u_{i\,j-1}^{k}}{2\Delta y}.$$

The whole algorithm is:

2.1 For every $(i = 1,\ldots,n-1)$ find $u_{i\,j}^{k+1}$

$$-0.5u_{i\,j-1}^{k+1} + u_{i\,j}^{k+1}(1+\alpha) - 0.5u_{i\,j+1}^{k+1} = r_{i\,j}, \qquad (j = 1,\ldots,m-1)$$

where $\alpha = \Delta y^2/(v\,\Delta t)$, $\quad r_{ij} = \Delta t\,\alpha f_{ij}^{k+1/2} + 0.5\,(u_{i\,j-1}^{k} - 2u_{i\,j}^{k} + u_{i\,j+1}^{k}) + \alpha\,u_{i\,j}^{k}$.

Boundary conditions.

Lower$(j=0)$: $u_{i\,0}^{k+1} = r_{i\,0} = 0$,

Upper$(j=k)$: $-u_{i\,m-1}^{k+1} + u_{i\,m}^{k+1} = r_{i\,m} = 0$.

In matrix form

$$\begin{pmatrix} 1 & 0 & 0 & 0 & 0 & 0 & 0 \\ -0.5 & 1+\alpha & -0.5 & 0 & 0 & 0 & 0 \\ 0 & -0.5 & 1+\alpha & -0.5 & 0 & 0 & 0 \\ 0 & 0 & -0.5 & 1+\alpha & -0.5 & 0 & 0 \\ 0 & 0 & 0 & -0.5 & 1+\alpha & -0.5 & 0 \\ 0 & 0 & 0 & 0 & -0.5 & 1+\alpha & -0.5 \\ 0 & 0 & 0 & 0 & 0 & -1 & 1 \end{pmatrix} \begin{pmatrix} u_{i\,0}^{k+1} \\ u_{i\,1}^{k+1} \\ u_{i\,3}^{k+1} \\ \ldots \\ u_{i\,m-2}^{k+1} \\ u_{i\,m-1}^{k+1} \\ u_{i\,m}^{k+1} \end{pmatrix} = \begin{pmatrix} r_{i\,0} \\ r_{i\,1} \\ r_{i\,2} \\ \ldots \\ r_{i\,m-2} \\ r_{i\,m-1} \\ r_{i\,m} \end{pmatrix}$$

Then for $(j = 0,\ldots,m)$

Left boundary $(i = 0)$: $u_{0\,j}^{k+1} = 0$,

Right boundary $(i = n)$: $u_{n\,j}^{k+1} = 0$.

2.2 For every $(i = 1, n-1)$ find

$v_{i\,0}^{k+1} = 0$,

$$v_{i\,j+1}^{k+1} = v_{i\,j}^{k+1} - 0{,}5\left\{\frac{u_{i+1\,j+1}^{k+1} - u_{i-1\,j+1}^{k+1}}{2\Delta x} + \frac{u_{i+1\,j}^{k+1} - u_{i-1\,j}^{k+1}}{2\Delta x}\right\}\Delta y, \qquad (j = 0,\ldots,m-1)$$

Then for $(j = 0,\ldots,m)$

Left boundary $(i = 0)$: $v_{0\,j}^{k+1} = 0$,

Right boundary $(i = n)$: $v_{n\,j}^{k+1} = 0$.

3. Results and discussion

The key role in appearance of Schlichting vortices plays thickness of the viscous acoustic boundary layer $\delta = (2\nu / \omega)^{0.5}$. After introducing dimensionless coordinates $y^* = y/\delta$, $x^* = x/L$ the mean longitudinal velocity profile across region with infinite width is [3]

$$\bar{u}_R = -\frac{U_0^2}{4c_0} \sin 2\pi\left(x^* + \frac{1}{2}\right)\left\{\frac{1}{2}e^{-2y^*} + \left(1 - y^*\right)e^{-y^*} \cos y^* + \left(4 + y^*\right)e^{-y^*} \sin y^* - \frac{3}{2}\right\},$$

where $c_0 = \lambda / T$ is sony velocity in standing wave.

In figures 1,3,5,7 the comparison between computed $\bar{u}^* = \bar{u} \Big/ \left(\frac{3}{16}\frac{U_0^2}{c_0}\right)$ and etalon

$\bar{u}_R^* = \bar{u}_R \Big/ \left(\frac{3}{16}\frac{U_0^2}{c_0}\right)$ dimensionless mean velocities at $x^* = 0.75$ is presented. At $H = 3\delta$, $H = 4\delta$ computed profiles are far from etalon. At $H = 6\delta$, $H = 8\delta$ computed profiles can be accepted.

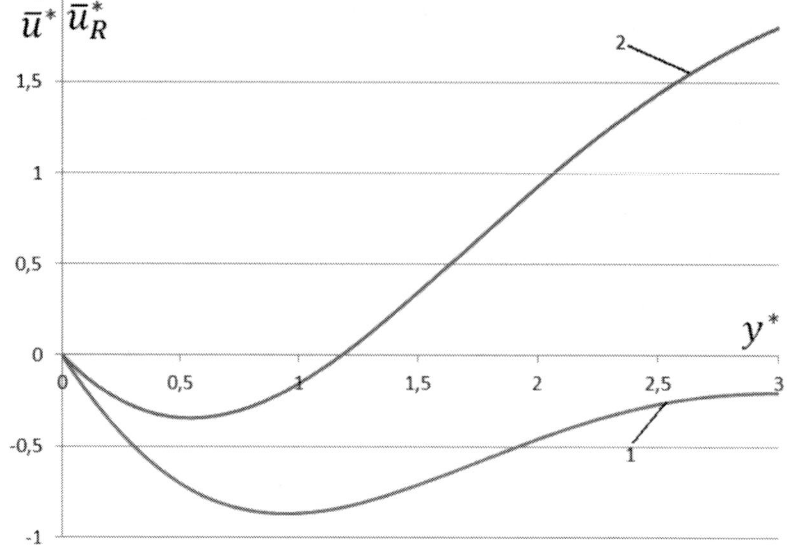

Figure 1 Mean velocity profiles at $H = 3\delta$. $1-\bar{u}^*$, $2-\bar{u}_R^*$

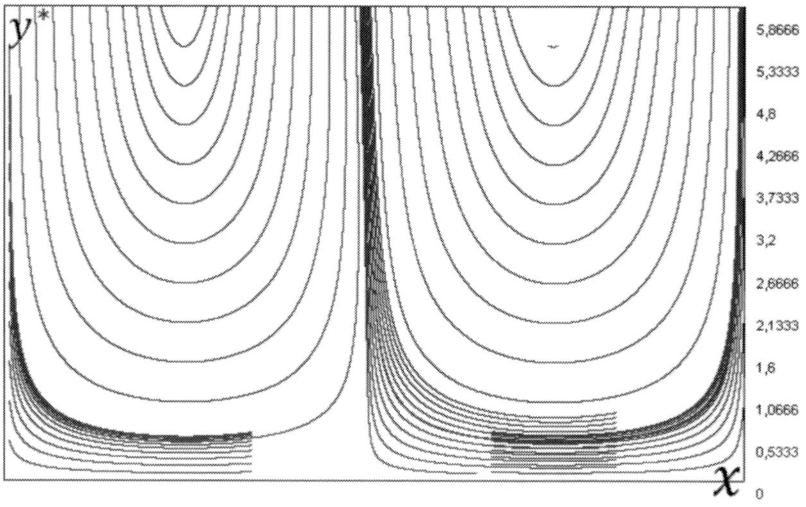

Figure 2 Mean velocity streamlines at $H = 3\delta$

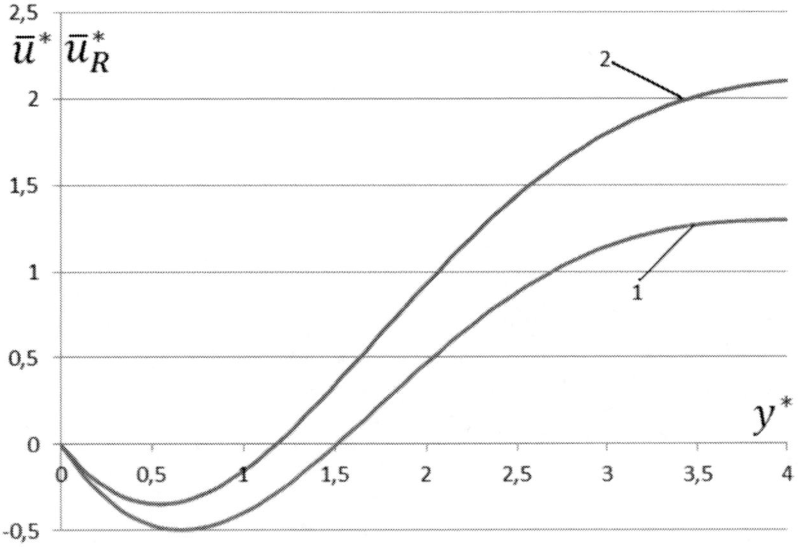

Figure 3 Mean velocity profiles at $H = 4\delta$. $1-\bar{u}^*$, $2-\bar{u}_R^*$

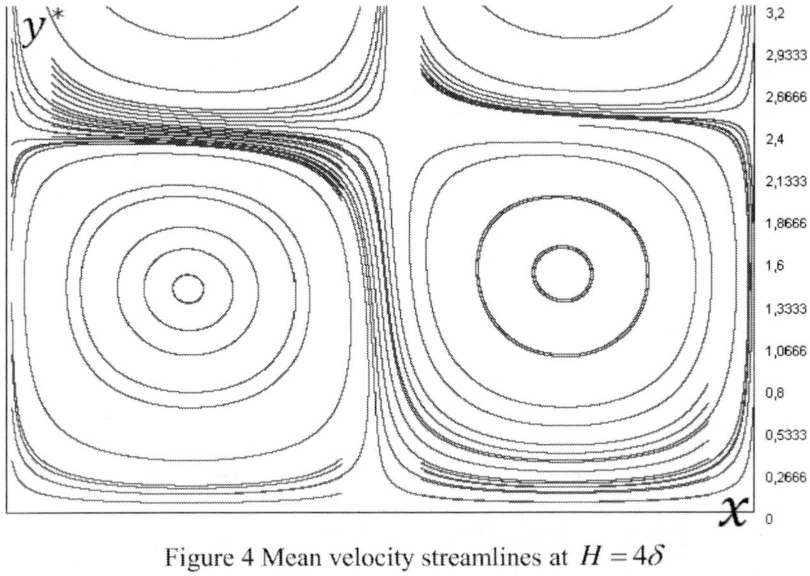

Figure 4 Mean velocity streamlines at $H = 4\delta$

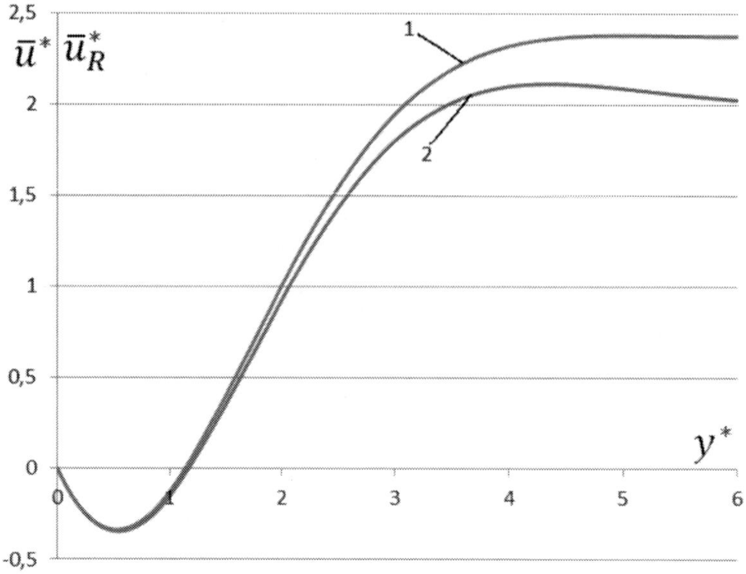

Figure 5 Mean velocity profiles at $H = 6\delta$. $1-\bar{u}^*$, $2-\bar{u}_R^*$

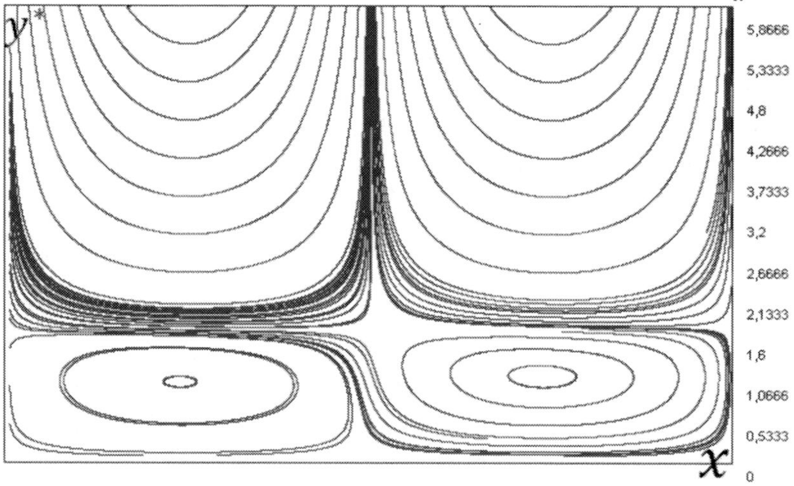

Figure 6 Mean velocity streamlines at $H = 6\delta$

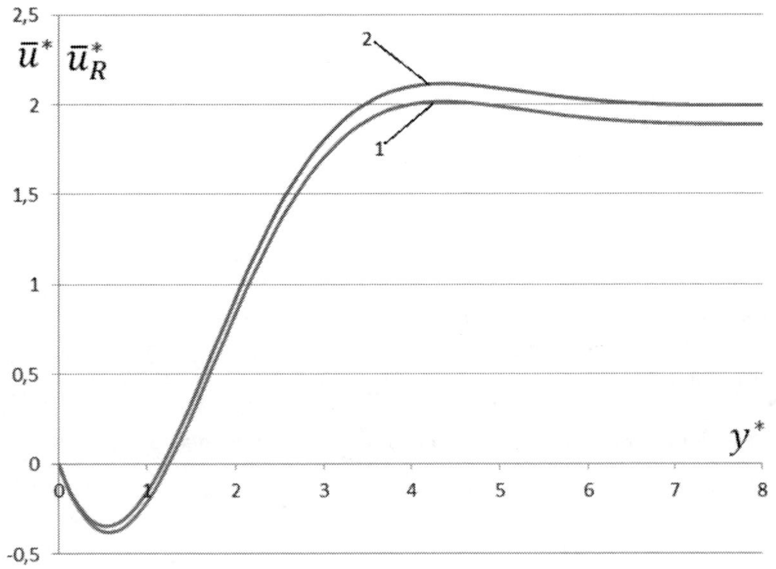

Figure 7 Mean velocity profiles at $H = 8\delta$. $1 - \bar{u}^*$, $2 - \bar{u}_R^*$

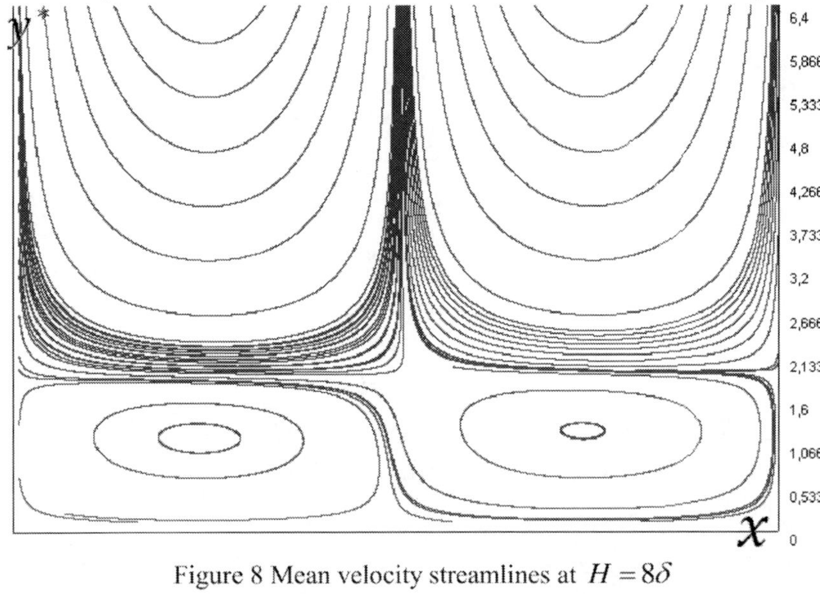

Figure 8 Mean velocity streamlines at $H = 8\delta$

In figures 2,4,6,8 the stream lines of acoustic streaming are depicted. In Figure 2 $(H = 3\delta)$ no Schlichting vortices are observed. In figures 4, 6, 8 one can see both Rayleigh and Schlichting vortices. The size of Schlichting vortex is near to ($L/2,\ 2\delta$). It seems that the condition for appearing the Schlichting vortices is $H \geq 4\delta$. Good approximation of acoustic streaming velocity is achieved at $H \geq 8\delta$. This means that computational region must be enough wide to resolve properly the Schlichting vortices.

Acknowledgments
This work was funded by the subsidy of the Russian Government to support the Program of Competitive Growth of Kazan Federal University among World's Leading Academic Centers.

References

[1] Aktas M K and Farouk B 2004 Numerical simulation of acoustic streaming generated by finite-amplitude resonant oscillations in an enclosure *J.Acoust. Soc. Am.* **116** 2811-2831
[2] Tajik B, Abbassi A, Saffar-Avval M, Abdullah A and Mohammad-Abadi H 2012 Numerical simulation of acoustic streaming for nonlinear standing ultrasonic wave in water inside axisymmetric enclosure *Engineering Applications of Complutational Fluid Mechanics* **6** 366-382
[3] Zarembo L K and Krasilnikov V A 1966 *Vvedenie v nelinejnuyu akustiku* (Moscow: "Nauka")

Journal of Physics: Conference Series **567** (2014) 012018 doi:10.1088/1742-6596/567/1/012018

Impact of Basset force on threshold values of particle drag coefficient and density parameter in standing sinusoidal wave

D A Gubaidullin[1,2], P P Osipov[1,2] and A N Zakirov[1,2]

[1]Institute of Mechanics and Engineering, Kazan Science Center, Russian Academy of Sciences, 2/31 Lobachevsky str., Kazan 420111, Russia
[2]Kazan (Volga Region) Federal University, 18 Kremlyovskaya str., Kazan 420008, Russia

E-mail: petro300@rambler.ru

Abstract. A one-dimensional drift of spherical particle in standing sinusoidal wave is studied numerically. The impact of stationary and non-stationary forces of viscous drag, as well as Archimedes, added masses and Basset forces on particle drift direction is investigated. For various Reynolds and Strouhal numbers the dependencies of the threshold particle drag coefficient on density parameter have been found. These dependencies show that with increasing Reynolds and Strouhal numbers the threshold value of the squared drag coefficient decreases markedly. Impact of Basset force on threshold values is especially strong for low-density particles.

1. Introduction

The well-known model of particle dynamic in viscous fluid is used [1-2]. Recently a new approach to investigation of particle drift in sinusoidal standing wave was developed in [3]. The drift of compressible particle in compressible fluid at low Reynolds number taking into account Stokes force, added mass, Archimedes forces and surface tension was investigated analytically. Small parameter method and Van-der-Pole method were used. Suggested approach is based on decomposition of particle motion into slow and fast parts (evolution and oscillation). Obtained formula for drift acceleration was laid to the base of diagram for drift direction. This diagram covers all possible frequencies of standing wave, and all relative particle densities. Diagram shows existence of threshold curves, dividing the drift regimes into several groups. In each group the drift of particle is directed to the nodes (or antinodes) of standing wave. In [4] the drift of incompressible inclusion at different Reynolds and Strouhal numbers under influence of viscosity, added mass and Archimedes forces was investigated numerically. The threshold curves for different Reynolds and Strouhal numbers were computed. On the base of the threshold curves of squared drag coefficient $\mu_p^2 = \{1+[\omega\tau_v\,(1+0.5\,\rho_1/\rho_2)]^2\}^{-1}$, where $\tau_v = \dfrac{2}{9}\,r^2\dfrac{\rho_2}{\mu}$ relaxation time, the diagram of particle drift direction in coordinates $\{D = 3\rho_1/(\rho_1+2\rho_2),\ \mu_p^2\}$ was suggested. It was shown that with increasing Reynolds and Strouhal numbers the threshold values of squared drag coefficient decrease significantly. In [5] the direction of drift of incompressible particle in acoustic resonator with periodic weak shock wave was studied numerically. Simulations revealed the existence of threshold value of particle radius which divides possible drift regimes in two groups with opposite drift directions.

2. Statement of the problem.

Dimensionless Cauchy problem for dynamics of particle is

$$\frac{dx_2}{dt}=\frac{1}{Sr}v_2,$$

$$\frac{dv_2}{dt}=-\frac{3D}{K}\left(1+\frac{1}{6}\sqrt{Re}+\frac{1}{60}Re\right)(v_2-v_1)+D\left(\frac{\partial v_1}{\partial t}+\frac{v_1}{Sr}\frac{\partial v_1}{\partial x}\right)+\frac{3D}{\sqrt{K}}\frac{1}{\sqrt{\pi}}\int_0^t\left(\frac{\partial v_1}{\partial \theta}+\frac{v_2}{Sr}\frac{\partial v_1}{\partial x}-\frac{dv_2}{d\theta}\right)\frac{d\theta}{\sqrt{t-\theta}},$$

initial conditions: $x_2(0)=0.75, \quad v_2(0)=0$.

Here $v_1(x,t)=sin(\pi x)sint$, $Sr=\omega L/U_0$ – Strouhal number, ω, L, U_0 - frequency, half of wave length and its amplitude respectively, $K=\omega r^2\rho_1/\mu$ - particles inertia criterion, μ -fluid viscosity, $D=3\rho_1/(\rho_1+2\rho_2)$ - density parameter, r - radius of particle, v_1 v_2 - velocities of fluid and particle, $Re=\rho_1|v_1-v_2|\frac{2r}{\mu}$ – relative Reynolds number, ρ_1, ρ_2 - densities of fluid and particle.

3. Computing Basset force.

Basset force is expressed via Duhamel integral

$$I\langle f(t)\rangle=\frac{1}{\sqrt{\pi}}\int_0^t f(\theta)\frac{d\theta}{\sqrt{t-\theta}}, \qquad \text{where} \quad f(\theta)=\frac{\partial v_1}{\partial \theta}+\frac{v_2}{Sr}\frac{\partial v_1}{\partial x}-\frac{dv_2}{d\theta}.$$

Computing Basset force becomes slow with time increase. Therefore the Duhamel integral was computed using parallel programming with help of MPI library. Parallel computations were performed on three quad core computers connected by LAN. On each time level $t_n=n\Delta t$ integration region was divided by the number of processors $n_p=12$

$$I^n=-\frac{2}{\sqrt{\pi}}\int_{k_0\Delta t}^{k_1\Delta t}f(\theta)\,d\sqrt{t_n-\theta}-\frac{2}{\sqrt{\pi}}\int_{k_1\Delta t}^{k_2\Delta t}f(\theta)\,d\sqrt{t_n-\theta}-\ldots-\frac{2}{\sqrt{\pi}}\int_{(k_{n_p}-1)\Delta t}^{t_n}f(\theta)\,d\sqrt{t_n-\theta},$$

where $k_i=\left[n/n_p\right]i \qquad (i=0,n_p-1), \quad t_n=k_{n_p}\Delta t=n\Delta t$.

The acceleration of computation by factor 9 was observed.

4. Impact of Basset force on direction of particle drift.

In diagram all possible particle drift directions are divided by curves of threshold drag coefficient as well as by vertical line of threshold density parameter. The dependencies of threshold values of squared drag coefficient $\mu_p^2=\{1+[\omega\tau_v(1+0.5\rho_1/\rho_2)]^2\}^{-1}$ on density parameter D are determined during simulations. In figure 1,2 the threshold values of μ_p^2 and D for different Strouhal numbers in non-Stokes regime $Re_0=10$ with and without presence of Basset force are depicted. Without Basset force there is a single threshold value of the density parameter for all Strouhal numbers. With Basset force, in contrary, there are different threshold values of density parameter D, which depend on Strouhal number. The higher the Strouhal number, the more the threshold density parameter.

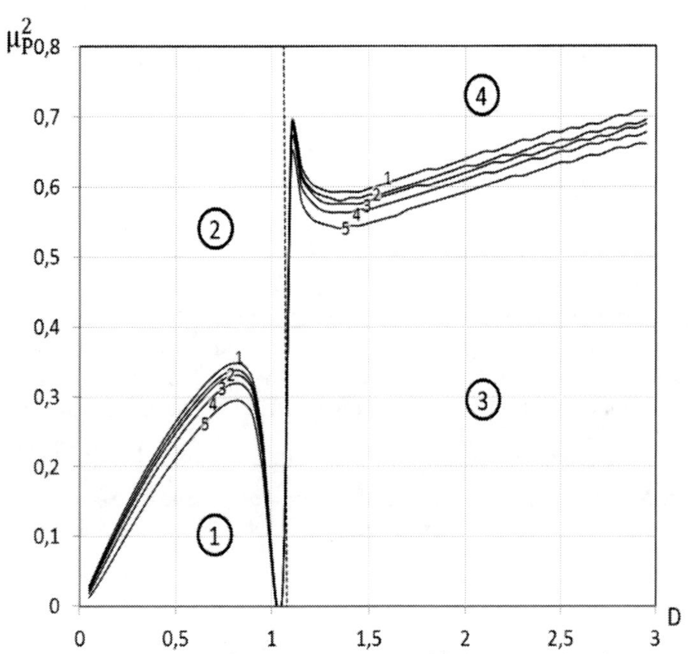

Figure 1. Threshold values of squared drag coefficient and density parameter without presence of Basset force.
Re_0=10; Sr=10(1), 20(2), 30(3), 40(4), 50(5).

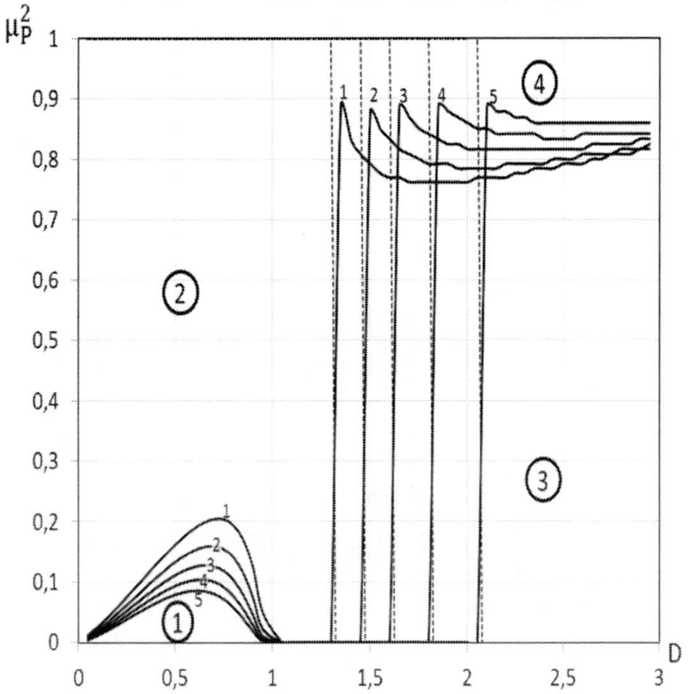

Figure 2. Threshold values of squared drag coefficient and density parameter with presence of Basset force.
Re_0=10; Sr=10(1), 20(2), 30(3), 40(4), 50(5).

Each curve divides the diagram into 4 parts. In parts 2, 3 the particle drift is directed to nodes of standing wave, in parts 1,4 –to antinodes. Calculations show that Basset force has a particularly strong influence on the threshold curves for low density particles. By increasing the Strouhal number, the threshold value μ_p^2 for dense particles decreases and the threshold value of the density parameter D increases.

5. Conclusion.

The threshold values of the squared drag coefficient and density parameter for various Reynolds and Strouhal numbers were computed. The influence of Basset force on the threshold values of the squared drag coefficient and parameter of density has been investigated. It was shown that presence of the Basset force has a particularly strong influence on the threshold curves for low density particles. Increase of the Strouhal number reduces threshold drag coefficient and increases density parameter.

Acknowledgments
This work was funded by the subsidy of the Russian Government to support the Program of Competitive Growth of Kazan Federal University among World's Leading Academic Centers.

References

[1] Nigmatulin R I 1991 *Dynamics of Multiphase Media* (New York: Hemisphere)
[2] Ganiev R F, Ukrainskyi L 2008 *Nonlinear wave mechanics and technology* (Moscow: press. R & C Dynamics)
[3] Gubaidullin D A, Osipov P P 2011 *J. of Eng. Phys. and Thermophys.* 84 255
[4] Gubaidullin D A, Ossipov P P 2013 *J. of Eng. Phys. and Thermophys.* 86 51
[5] Gubaidullin D A, Ossipov P P 2013 *J. Applied Math. and Computation* 219 4535

Acoustic waves in multifractional gas mixture with the inclusion of different materials and dimensions without Phase Transformations

D A Gubaidullin[1,2], E A Teregulova [1,2]

[1] Institute of Mechanics and Engineering, Kazan Science Center, Russian Academy of Sciences, 2/31 Lobachevsky str., Kazan 420111, Russia
[2] Kazan (Volga Region) Federal University, 18 Kremlyovskaya str., Kazan 420111, Russia

E-Mail: teregulova@inbox.ru

Abstract. The propagation of acoustic waves in mixtures of gas and particle fractions of different materials and sizes is studied. A mathematical model is presented, the dispersion equation is obtained, dispersion curves are calculated. The influence of the particle size and the parameters of the dispersed phase for multifractional gas mixture with ice particles, aluminum and sand on dissipation and dispersion of sound waves is analyzed. A comparison with experiment is conducted.

1. Introduction

The study of acoustics and wave dynamics of multiphase media is of considerable theoretical and practical interest in view of the wide spread of such media in nature and their use in practice [1]. Basic model s of wave dynamics of dispersed media and a number of results in this area are presented in [1]. Work [2] is devoted to the problems of the investigation of two-phase flows with solid particles, droplets and bubbles. In monography [3] a brief overview of the results on the study of acoustic disturbances in monodisperse gas mixtures without phase transformations is given. Influence of polydisperse composition of the gas mixture on the distribution of monochromatic disturbances in one-component mixtures of gas and particles, or vapor with droplets is studied in [4]. Peculiarities of the propagation of monochromatic waves in two-component polydisperse mixtures of gas and vapor and liquid droplets are investigated in [5]. Propagation of spherical and cylindrical waves of small amplitude in polydispersed fogs with phase transformations is considered in [6]. A general dispersion relation of the wave number to the oscillation frequency and thermal properties of the phases is obtained. The anomalous effect of nonmonotonic dependence of dissipation of weak harmonic and pulse disturbances on the mass concentration of droplets in monodisperse aerosols with heat and mass transfer is studied in [7]. Fairly complete presentation of the propagation linear theory of flat disturbances in mono-and polydisperse two-phase gas-vapor-liquid droplet mixtures is given in [8]. The propagation of acoustic waves of different geometry in two- fraction gas mixtures with particles

[1] To whom any correspondence should be addressed.

of different sizes and materials, without taking phase transformations into account is studied in [9]. Peculiarities of plane, cylindrical and spherical waves of small amplitude in the gas-vapor-droplet mixtures with solid particles are analyzed in [10], [11], [12]. In this paper the propagation of plane, cylindrical, and spherical waves of small amplitude in multifractional mixtures of gas with an arbitrary number of solid particles of different sizes and substances with significantly different thermal properties is studied.

2. Basic equations and dispersion equation.

Similarly to [8], the linearized system of equations of disturbed motion of multifractional gas mixture with solid particles of different materials and sizes, in a coordinate system relative to which the undisturbed medium is at rest, can be written as

$$\frac{\partial\rho_1'}{\partial t}+\rho_{10}\left(\frac{\partial v_1'}{\partial r}+\theta\frac{v_1'}{r}\right)=0,\quad \frac{\partial\rho_j'}{\partial t}+\rho_{j0}\left(\frac{\partial v_j'}{\partial r}+\theta\frac{v_j'}{r}\right)=0,\quad \left(j=\overline{2,N}\right)$$

$$\rho_{10}\frac{\partial v_1'}{\partial t}+\frac{\partial p_1'}{\partial r}+\sum_{j=2}^{N}n_{0j}\left(6\pi a_j\mu_1\left(v_1'-v_j'\right)+6a_j^2\sqrt{\pi\mu_1\rho_1^\circ}\int_{-\infty}^{t}\frac{\partial}{\partial\tau}\left(v_1'-v_j'\right)\frac{d\tau}{\sqrt{t-\tau}}\right)=0,$$

$$\rho_{j0}\frac{\partial v_j'}{\partial t}=n_{0j}\left(6\pi a_j\mu_1\left(v_1'-v_j'\right)+6a_j^2\sqrt{\pi\mu_1\rho_1^\circ}\int_{-\infty}^{t}\frac{\partial}{\partial\tau}\left(v_1'-v_j'\right)\frac{d\tau}{\sqrt{t-\tau}}\right),\quad \left(j=\overline{2,N}\right)$$

$$\rho_{10}c_{p1}\frac{\partial T_1'}{\partial t}=\alpha_{10}\frac{\partial p_1'}{\partial t}-\sum_{j=2}^{N}2\pi n_{0j}a_j\lambda_1 Nu_{1j}\left(T_1'-T_{\Sigma j}'\right),\tag{1}$$

$$\rho_{j0}c_j\frac{\partial T_j'}{\partial t}=-2\pi n_{0j}a_j\lambda_j Nu_j\left(T_j'-T_{\Sigma j}'\right),\quad \left(j=\overline{2,N}\right)$$

$$\lambda_1 Nu_{1j}\left(T_1'-T_{\Sigma j}'\right)+\lambda_j Nu_j\left(T_j'-T_{\Sigma j}'\right)=0,\quad \left(j=\overline{2,N}\right)$$

$$p_1'=\frac{C_1^2}{\gamma_1\alpha_{10}}\rho_1'+\frac{p_0}{T_0}T_1'.$$

The system of equations (1) for the parameter $\theta=0$ describes plane waves in a Cartesian coordinate system , when $\theta=1$ - cylindrical waves in a cylindrical coordinate system, when $\theta=2$ - spherical waves in a spherical coordinate system. Here and further, variables with subscript 1 refer to the carrier phase, with indexes $\overline{2,N}$ to the particles of the dispersed phase of the radius $a_j,\left(j=\overline{2,N}\right)$. Index 0 corresponds to the initial undisturbed state. Dashes above are used to denote disturbed parameters. Here and further ρ- reduced density, ρ°- true density , v - speed , α- volume content , p - pressure , μ_1- coefficient of the dynamic viscosity of the carrier medium, T_1 - the temperature of the carrier phase , $T_{\Sigma j}$ - the temperature in the surface Σ- layer of a particle of j - type , T_j- temperature of solid particles of j- type , Nu_{1j} - the dimensionless (Nusselt number) coefficient of the carrier phase heat transfer with the border of the interface of j- type particles , Nu_j - the dimensionless (Nusselt number) coefficient of the heat transfer of j- type particles with the border of the interface, λ - coefficient of thermal conductivity.

The system of equations (1) is closed and can be used to investigate the propagation of acoustic disturbances in polydisperse mixtures of gas and solid particles of different thermal properties and dimensions in flat , spherical and cylindrical cases.

Let us introduce into the system of equations (1) the phase velocity potentials and explore the solutions of the resulting system of equations in the form of progressive waves for disturbances

$$\eta'=A_\eta\psi\tag{2}$$

where

$\psi = \exp\left[i(K_* r - \omega t)\right]$ - for flat disturbances

$\psi = H_0^{(1)}(K_* r)\exp(-i\omega t)$ - for cylindrical disturbances

$\psi = \dfrac{1}{r}\exp\left[i(K_* r - \omega t)\right]$ - for spherical disturbances

$$K_* = K + iK_{**},\ C_p = \frac{\omega}{K},\ \sigma = 2\pi\frac{K_{**}}{K}.$$

Here A_η - the oscillation amplitude of parameters disturbances , K_* - the complex wave number , i - the imaginary unit, C_p - the phase velocity, K_{**} -the linear attenuation coefficient, σ - the attenuation decrement on the wave length, $H_0^{(1)}(z)$ - Hankel function , which is a combination of Bessel functions of the first and second kind of order zero $J_0(z)$ and $Y_0(z)$ $\left(H_0^{(1)}(z) = J_0(z) + iY_0(z)\right)$. Substituting the velocity potentials and solutions of the form (2) in the system of equations (1), we obtain the following system of linear algebraic equations and from conditions of the existence of nontrivial solution of a system of linear algebraic equations the following dispersion equation is obtained

$$\left(\frac{C_1 K_*}{\omega}\right)^2 = V(\omega)D(\omega), \qquad (3)$$

where

$$V(\omega) = 1 + \sum_{j=2}^{N}\frac{m_j}{1 - i\omega\tau_{vj}^*},\ \overline{\tau}_{Tj}^* = \tau_{Tj}^* + \frac{m_j c_j}{c_{p1}}\tau_{T1j}^*,$$

$$D(\omega) = 1 + (\gamma_1 - 1)\frac{\sum_{j=2}^{N}\frac{m_j c_j}{c_{p1}}\frac{1}{1 - i\omega\overline{\tau}_{Tj}^*}}{1 + \sum_{j=2}^{N}\frac{m_j c_j}{c_{p1}}\frac{1}{1 - i\omega\overline{\tau}_{Tj}^*}}.$$

3. Results.

As an example the mixture of air and particles of aluminum of the radius $r_a = 10^{-6}$ m , sand of the radius $r_s = 8\cdot 10^{-6}$ m and ice of the radius $r_i = 10^{-5}$ m. Fig . 1,2 illustrate the effect of inclusions mass content on the dependences of the relative speed of sound and attenuation decrement on the wavelength on the dimensionless oscillation frequency $\omega\tau_{vs}$ respectively. The calculated curves are built using the dispersion relation (10). The graphs show that with increasing mass content of particles the sound speed dispersion and dissipation of the waves increase. Accounting of the threefraction composition and difference of thermal parameters of fractions lead to the appearance of characteristic inflexions for the dependence of the relative speed of sound in the frequency field inversely proportional to the characteristic relaxation times of the phase velocities τ_{va}, τ_{vb} and τ_{vs} (Fig. 3) . As shown in Fig. 4 , the difference between the sizes of inclusions and thermal parameters of particles fractions leads to appearance of three maxima for the dependence of attenuation decrement on the wavelength at the characteristic values , $\omega\tau_{va}$, $\omega\tau_{vl}$ and $\omega\tau_{vs} = 1$.

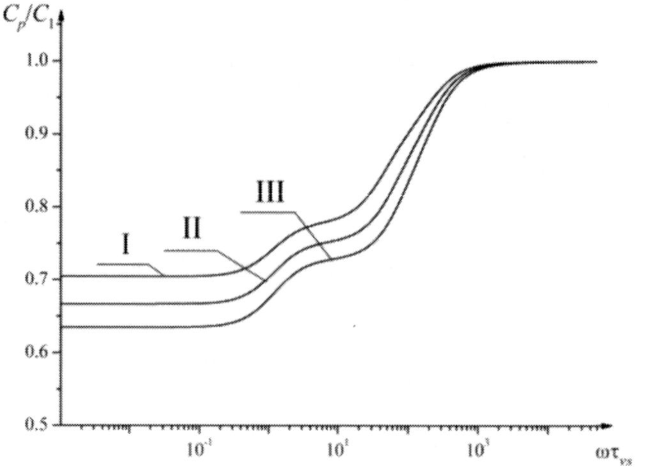

Figure 1. Dependence of the relative speed of sound on the dimensionless oscillation frequency for a mixture of air with particles of ice , aluminum and sand with different mass content of dispersed phase (I: m = 0.3, II: m = 0.5, III: m = 0.7)

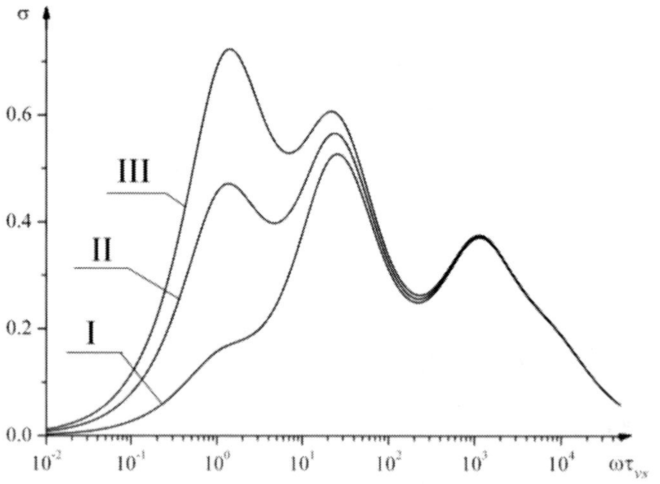

Figure 2 .The attenuation decrement dependence on the wavelength on the dimensionless oscillation frequency for a mixture of air with particles of ice, aluminum and sand with different mass content of dispersed phase (I: m = 0.3, II: m = 0.5, III: m = 0.7).

4. Comparison with experiment.

An experimental investigation of attenuation and dispersion of sound waves in a mixture of air with particles of aluminum of four different radii $r_1 = 5 \cdot 10^{-6}$ m, $n_1 = 9.81 \cdot 10^{9}$ m^{-1}, $r_2 = 7.5 \cdot 10^{-6}$ m,

$n_2 = 8.49 \cdot 10^9\,\mathrm{m}^{-1}$, $r_3 = 10^{-5}\,\mathrm{m}$, $n_3 = 5.99 \cdot 10^9\,\mathrm{m}^{-1}$, $r_4 = 1.5 \cdot 10^{-5}\,\mathrm{m}$, $n_4 = 1.095 \cdot 10^9\,\mathrm{m}^{-1}$ in the frequency range 500 - 13600Gts is conducted in [13]. Figure 3 shows a comparison of theoretical and experimental data for the attenuation coefficient on the wavelength per unit fraction mass $\bar{\alpha} = C_1 K_{**} \Big/ \left(\omega \sum_{j=2,N} \frac{n_{0j} g_{0j}}{\rho_1} \right)$ [13].

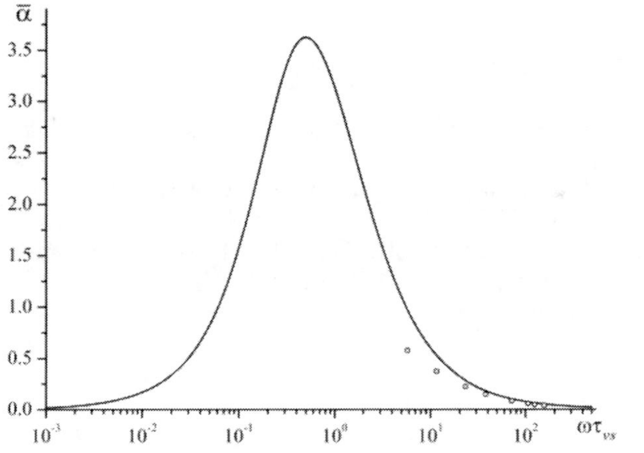

Figure 3. Comparison of theory with experimental data for dependencies of the attenuation coefficient on the wavelength per unit fraction mass $\bar{\alpha}$ on the dimensionless frequency of disturbances $\omega\tau_{vs}$.

The graph shows that in the considered frequency range experimental and theoretical values agree satisfactorily. However , the experiment of Zink and Delsasso was conducted in a frequency range that does not allow to show the presence of the maximum attenuation on the wavelength and maximum dispersion .

5. Conclusion.

This paper presents a closed system of linear differential equations of motion for multifractional mixture of gas and solid particles of different sizes and thermal properties. A dispersion equation, determining the propagation of plane, spherical and cylindrical disturbances of small amplitude derived. The dispersion curves are calculated. The influence of the parameters of the dispersed phase for three-fraction gas mixture with aluminum particles, carbon black, and ice on the dissipation and dispersion of sound waves is analyzed .It is found that the difference of thermal properties and inclusions sizes significantly affects on the dynamics of weak waves in multifractional gas mixtures and it must be considered in the development of methods of acoustic diagnostics of considered environments. Good agreement of theoretical and experimental data is shown.

Acknowledgments
This work was funded by the subsidy of the Russian Government to support the Program of Competitive Growth of Kazan Federal University among World's Leading Academic Centers.

References

[1] Nigmatulin R I 1991 Dynamics of Multiphase Media vol 1,2 (Washington: Hemisphere)
[2] Varaksin A Yu 2013 Fluid dynamics and thermal physics of two-phase flows: Problems and achievements *High Temperature* **51(3)** 421-455
[3] Temkin S 2005 *Suspension acoustics: An introduction to the physics of suspension* (Cambridge: Cambridge University Press) p 398
[4] Gumerov N A and Ivandaev A I 1988 Sound propagation in polydispersed gas suspensions *Journal of Applied Mechanics and Technical Physics* **5** 115-124.
[5] Gubaidullin D A and Ivandaev A I 1993 Effect of polydispersion on sound propagation in gas mixtures with vapor and liquid drops *Journal of Applied Mechanics and Technical Physics* **4** 75-83
[6] Gubaidullin D A 2003 Spherical and Cylindrical Low-Amplitude Waves in Polydisperse Fogs with Phase Transitions *Fluid Dynamics* **5** 85-94
[7] Nigmatulin R I, Ivandaev A I and Gubaidullin D A 1991 Effect of the non-monotonic dependence of the dissipation of sound on the concentration of drops in a suspension in gas *Dokl. USSR Academy of Sciences* **316(3)** 601-605
[8] Gubaidulln D A 1998 *Dynamics of twophase gas-vapor-droplet media* (Kazan: Kazan mathematical society) p 153
[9] Gubaidullin D A, Nikiforov A A and Utkina E A 2009 Propagation of acoustic waves to two-fractional gas mixtures with particles of different materials and sizes *Izvestiya vysshih uchebnykh zaavedenii. Problemi Energetiki* **1-2** 25-33.
[10] Gubaidullin D A, Nikiforov A A and Utkina E A 2011 Effect of the phase transformations on propagation of acoustic waves in two-fraction mixtures of gas with vapor, droplets and solid particles of different materials and sizes *High Temperature* **49(6)** 942-947
[11] Gubaidullin D A, Nikiforov A A and Utkina E A 2011 Acoustic waves in two-fraction mixtures of gas with vapor, droplets and solid particles of different materials and sizes in the presence of phase transitions *Fluid Dynamics* **1** 95-103
[12] Nikiforov A A, Utkina E A and Gafiyatov R N 2011 Acoustic disturbances in vapor-gas-liquid systems *Vestnik of Lobachevsky State University of Nizhni Novgorod* **4(3)** 1017-1018
[13] Zink J W and Delsasso L P 1958 Attenuation and Dispersion of Sound by Solid Particles Suspended in a Gas *J. Acoust. Soc. Am.* **30** 765-771

Acoustic waves in bubbly liquids for two kinds of gas bubbles with phase transitions

D A Gubaidullin[1,2] and R N Gafiyatov[1,2]

[1] Institute of Mechanics and Engineering, Kazan Science Center, Russian Academy of Sciences, 2/31 Lobachevsky str., Kazan, 420111, Russia
[2] Kazan (Volga Region) Federal University, 18 Kremlyovskaya str., Kazan, 420008, Russia

E-mail: gafiyatov@mail.ru

Abstract. The propagation of acoustic waves in a mixture of liquid with vapor-gas bubbles is studied. The dispersed phase consists of two fractions of the bubbles differing on gas structure and radii (a two-fractional mixture). The phase transitions take place at each of the fractions. The volume content of the bubbles is the low (about 1%). The dispersion relation, unified for plane, cylindrical and spherical waves is obtained. It is shown that a presence of the second fraction in the structure of a disperse phase of the bubbles changes significantly a dispersion and a dissipation of acoustic waves.

1. Introduction

A propagation of acoustic waves in a liquid with gas bubbles is studied intensively both theoretically and experimentally. Nigmatulin [1], Nakoryakov et al. [2], Temkin [3] considered various basic aspects of acoustics of multiphase flows. Hao and Prosperetti [4] obtained the results of study of the vapor bubble dynamics in liquid. Commander and Prosperetti [5] presented a model of the propagation of plane waves with small amplitude in a mixture of liquid and of gas bubbles. This model is adequate for the frequencies far from resonance at the volume content of a disperse phase of 1-2%. Nigmatulin et al. [6] showed the need of allowing for the effect of liquid compressibility for problems of acoustics of bubbly liquids. Shagapov [7] considered a problem of propagation of small disturbances in liquids with polydisperse bubbles. Nigmatulin et al. [8] studied the small radial oscillations of vapor-gas bubbles in liquid under the influence of an acoustic field. They showed that capillary effects and phase transitions lead to the new resonant frequency of small bubbles. This frequency is different from the Minnaert frequency. Vakhitova and Shagapov [9] investigated the propagation of small disturbances in two-component two-phase bubbling media with a three-temperature model. They found that heat and mass transfer rather than an interphase friction mainly determine dispersion. Gubaidullin and Nikiforov [10] obtained a dispersion relation describing a propagation of harmonic spherical and cylindrical disturbances in monodisperse mixture of a liquid with the vapor-gas bubbles. They showed strong dependence of an attenuation of pulse waves on a vapor concentration in bubbles. Gubaidullin, Nikiforov and Gafiyatov [11] investigated a propagation of acoustic waves in a mixture of liquid with vapor-gas bubbles and with insoluble gas bubbles taking into account the phase transitions only in the one fraction.

In the present paper, a propagation of a sound in the mixture of liquid with two kinds of vapor-gas bubbles with the phase transitions on the bubble surface is studied.

2. Dispersion relation

Linearized equations for one-dimensional disturbances in a two-fractional mixture of liquid with vapor-gas bubbles are obtained from the general equations of motion for two-phase mixtures [1]. Solving this system this dispersion relation is obtained

$$\left(\frac{K_*}{\omega}\right)^2=\frac{\alpha_{10}^2}{C_1^2}+\frac{\alpha_{a0}\rho_{10}}{p_0N_{Ra}}\left(1+\frac{H_{1a}+H_{1b}+H_{2b}\chi_1}{M_{3a}\chi_2/B_{1a}}\right)+\frac{\alpha_{b0}\rho_{10}}{p_0N_{Rb}}\left(1+\frac{H_{1a}+H_{1b}+H_{2a}\chi_3}{M_{3b}\chi_4/B_{1b}}\right) \quad (1)$$

Here

$$\chi_1=\frac{M_{4b}}{M_{3b}}-\frac{B_{1b}M_{4a}}{B_{1a}M_{3b}},\ \chi_2=\frac{H_{2a}B_{1a}}{M_{3a}}+\frac{H_{2b}B_{1b}}{M_{3b}},\ \chi_3=\frac{M_{4a}}{M_{3a}}-\frac{B_{1a}M_{4b}}{B_{1b}M_{3a}},$$

$$\chi_4=\frac{H_{2a}B_{1a}}{M_{3a}}+\frac{H_{2b}B_{1b}}{M_{3b}},$$

$$\delta=\frac{m_j^{\circ}-1}{1-k_{Vj0}},\ H_{1j}=\frac{m_j}{B_{1j}\tau_{T1j}}\left(B_{1j}M_{2j}+N_{1j}M_{4j}\right),\ H_{2j}=\frac{m_j}{B_{1j}\tau_{T1j}}\left(B_{1j}M_{1j}-N_{1j}M_{3j}\right),$$

$$M_{1j}=G_j-M_{2j}-\frac{L_{1j}N_{3j}}{L_{4j}-\delta N_{2j}},\ M_{2j}=\frac{N_{2j}L_{1j}}{N_{Rj}}\frac{1}{L_{4j}-\delta N_{2j}},$$

$$M_{3j}=\frac{N_{3j}\delta-L_{2j}N_{3j}}{L_{4j}-\delta N_{2j}}+L_{3j}+M_{4j},\ M_{4j}=\frac{L_{4j}-L_{2j}N_{2j}}{N_{Rj}(L_{4j}-\delta N_{2j})},$$

$$B_{1j}=\frac{c_1\tau_{T2j}}{c_{2j}\tau_{T1j}},\ N_{Rj}=-\frac{\rho_{10}^{\circ}}{p_0}\frac{(i\omega)(j_0)^2G_{Rj}}{3(G_{Rj}t_{Aj}+1)},\ G_{Rj}=\frac{1}{t_{Rj}}-i\omega,$$

$$t_{Rj}=\frac{(j_0)^{2^2}}{4\nu_1},\ t_A^j=\frac{j_0}{C_1(\alpha_{j0})^{1/3}},$$

$$N_{1j}=\frac{i\omega\tau_{T1j}}{2m_j}-1,\ N_{2j}=i\omega\tau_{T2j}-1,\ N_{3j}=k_{2j}(c_{2j}-R_j)-1+G_j,$$

$$L_{1j}=E_j(i\omega\tau_j-1),\ L_{2j}=-\frac{l_0k_{2j}}{(1-k_{Vj0})T_0}+\Delta R_j-L_{1j}(1+B_{1j}),$$

$$L_{3j}=1-G_j(1+B_{1j}),\ L_{4j}=L_{1j}+\Delta R_jN_{2j},\ k_{2j}=\frac{i\omega\tau_{T2j}}{c_{2j}},$$

$$m_j^{\circ}=\frac{\rho_{j0}^{\circ}}{\rho_{10}^{\circ}},\ m_j=\frac{\rho_{2j0}}{\rho_{10}},\ j=a,b.$$

Dispersion relation (1) (i.e. the function of the complex wave number K_* on the frequency ω) determines a propagation of acoustic plane, cylindrical and spherical waves in two-fractional mixtures of liquid with vapor-gas bubbles (different initial radii a_0, b_0, different initial volume contents α_{2a0}, α_{2b0}, and different thermal properties of fractions) with the interphase diffusion mass transfer in each fraction.

3. Results

Figure 1 shows the phase velocity and attenuation coefficient in water for the volume contents of the fraction of $\alpha_{2a} = \alpha_{2b} = 0.005$, for the radius of air bubbles of $a_0 = 2\cdot10^{-3}$ m, for the radius of helium bubbles (solid lines) of $b_0 = 10^{-3}$ m, calculated by dispersion relation (1). In unperturbed state a pressure of a two-fractional mixture is $p_0 = 0.1\,\text{MPa}$, a temperature is $T_0 = 327\,\text{K}$, a mass concentration of water vapor in bubbles is $k_{Va0} = k_{Vb0} = 0.1$. This also presents the results of calculations for the monodisperse mixtures of water with air bubbles ($a_0 = 2\cdot10^{-3}\,\text{m}, \alpha_2 = 0.01$, dashed lines) and helium bubbles ($a_0 = 10^{-3}\,\text{m}$, $\alpha_2 = 0.01$, dotted lines).

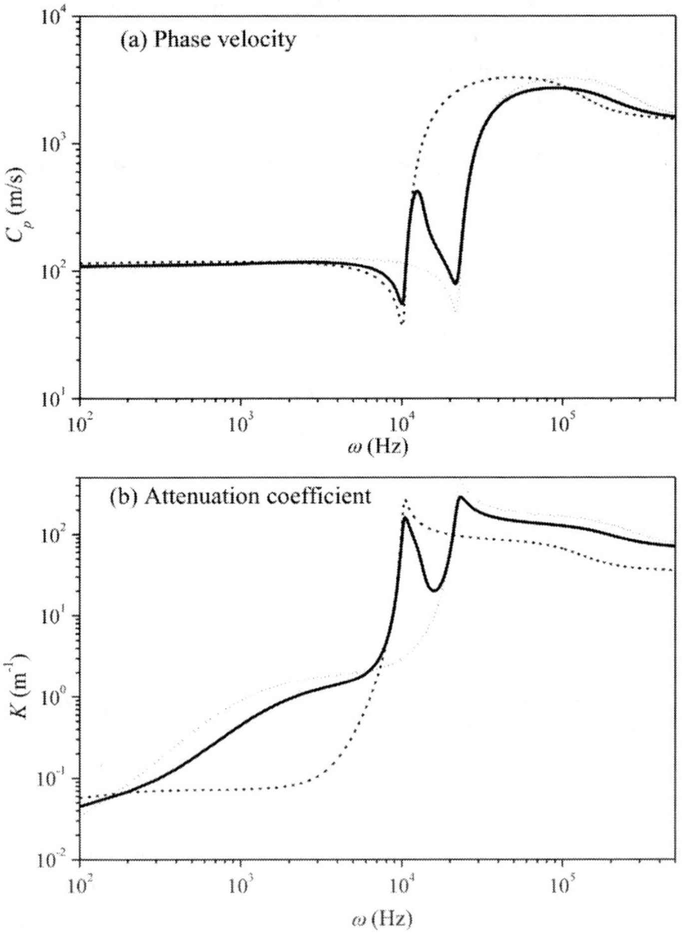

Figure 1. Frequency dependences of the phase velocity (a) and of the attenuation coefficient (b) for the mixture of water with: vapor–air and vapor–helium bubbles (solid line), vapor–air bubbles (dashed line), vapor–helium bubbles (dotted line).

The difference of gases in bubbles have small effect on the phase velocity value (figure 1(a)). However, the presence of two different bubble sizes in the mixture is important significantly. This leads to the occurrence of the two minimum points of the phase velocity curve in the near of the resonance frequencies of the oscillations of bubbles and to a characteristic inflection of a phase velocity curve (figure 1(a), solid line). In case of propagation of the harmonic waves in a liquid with bubbles of one radius, only one minimum of the phase velocity is observed (figure 1(a), dashed and dotted lines). The phase velocity values (figure 1(a), solid line) in low- and high-frequency areas are equal to the values for the monodisperse mixtures (figure 1(a), dashed and dotted lines), because the volume content of bubbles is the same in both cases ($\alpha_2 = 0.01$, $\alpha_{2a} = \alpha_{2b} = 0.005$).

The phase velocity and the attenuation coefficient have maximum points, corresponding to the two resonance frequencies (figure 1(b), solid lines). By equal volume content of bubbles ($\alpha_2 = 0.01$) the value of the attenuation coefficient in low-frequency area in two-fractional mixture of water with vapor–air and vapor–helium bubbles (figure 1(b), solid line) is more, than the value for mixture of water with vapor–air bubbles only (figure 1(b), dotted line).

4. Conclusions

The propagation of acoustic waves in two-fractional mixture of liquid with vapor-gas bubbles taking into account the phase transitions is studied. The dispersed phase consists of two fractions of the bubbles with different size and gas structure. The presence of two fractions of bubbles leads to occurrence of two maxima in a frequency dependence of attenuation coefficient, and this causes a characteristic inflection of a frequency dependence of phase velocity. It is shown that dispersion and dissipation of acoustic waves depends significantly on presence of the second fraction in a part of bubbles.

Acknowledgments

This work was funded by the subsidy of the Russian Government to support the Program of Competitive Growth of Kazan Federal University among World's Leading Academic Centers.

References

[1] Nigmatulin R I 1991 *Dynamics of Multiphase Media* (New York: Hemisphere)
[2] Nakoryakov V E, Pokusaev B G and Shreiber I R 1993 *Wave Propagation in Gas-Liquid Media* (New York: CRC Press)
[3] Temkin S 2005 *Suspension Acoustics: An Introduction to the Physics of Suspensions* (New York: Cambridge University Press)
[4] Hao Y and Prosperetti A 1999 *Physics of Fluids* **11** 2008
[5] Commander K W and Prosperetti A 1989 Journal of the Acoustical Society of America **85** 732
[6] Nigmatulin R I, Shagapov V Sh and Vakhitova N K 1989 *USSR Academy of Sciences Doklady* **304** 1077
[7] Shagapov V Sh 1977 *Journal of Applied Mechanics and Technical Physics* **18** 77
[8] Nigmatulin R I, Khabeev N S and Nagiev F B 1981 *International Journal Heat and Mass Transfer* **24** 1033
[9] Vakhitova N K and Shagapov V Sh 1984 *Journal of Applied Mechanics and Technical Physics* **25** 682
[10] Gubaidullin D A and Nikiforov A A 2010 *High Temperature* **48** 170
[11] Gubaidullin D A, Nikifirov A A and Gafiyatov R N 2012 *High Temperature* **50** 248

Oscillations of aerosols in the tubes near to resonances

D A Gubaidullin[1,2], R G Zaripov[1,2] and L A Tkachenko[1,2]

[1] Institute of Mechanics and Engineering, Kazan Science Center, Russian Academy of Sciences, 2/31 Lobachevsky str., Kazan 420111, Russia
[2] Kazan (Volga Region) Federal University, 18 Kremlyovskaya str., Kazan 420111, Russia

E-mail: luda_tkachenko@inbox.ru

Abstract. Experimental investigations longitudinal oscillations of a small-dispersed aerosol were carried out for tubes with various geometry on the end in a shock-free modes near to natural resonances. Di-ethyl-hexyl-sebacate was used as the working fluid to generate aerosol. Number concentration of drops for all experiments decreases with time scale and with growth of the excitation frequency. In the closed tube this process includes the coagulation and deposition of aerosol droplets on the tube walls. In the open tube except the coagulation and deposited of droplets on the walls of tube processes as unsteady discharge of aerosol from the open end of the tube. The time dependence of the numerical concentration of aerosol droplets is determined, and the effect of frequency and amplitude of oscillation on the process of coagulation and deposition of droplets is studied.

1. Introduction

In modern equipment, diverse inhomogeneous and multiphase media serve as the main working media. In so doing, numerous phenomena exhibit a wave and unsteady-state pattern. The investigation of wave processes in such media is an urgent problem in the dynamics and thermophysics of multiphase systems. Of the diversity of heterogeneous media, disperse flows may be identified which are relatively regular and represent a mixture of several phases, of which one consists of various inclusions (such as droplets, bubbles, solid particles), i.e., aerosols, fogs, bubble liquids, suspensions, and the like. Basic models of wave dynamics of multiphase media and some related results are given in monograph [1]. The oscillation processes in two-phase bounded media are treated in detail in [2, 3]. The dynamics of travelling waves in gas-droplet systems are dealt with in [4, 5]. Of special importance is the study of oscillation with highly nonlinear fronts of waves and coagulation of aerosols, because both are of interest from the standpoint of practical applications (intensification of the atomization of liquids with a view to cleaning waste gases in harmful production processes, deposition of aerosol in various technical devices, intensification of stirring and combustion, and the like). The acoustic coagulation of aerosols, which is accompanied by agglomeration of small particles or droplets under the effect of acoustic waves, is a well-known phenomenon treated in detail in the monographs [6-11] and papers [12-14] and in other publications. Widely employed is a system of the type of resonance tube, in which the excitation of the medium in the vicinity of resonance causes the emergence of

[1] To whom any correspondence should be addressed.

highly nonlinear pressure waves up to periodic shock waves at resonance. The investigations of nonlinear resonance oscillation of a homogeneous gas in tubes with different conditions at the ends are reviewed in [15]. It is noted that the resonant modes of oscillation are accompanied by a number of nonlinear effects such as the turbulization of flow, emergence of secondary flows and acoustothermal processes, generation of higher harmonics, formation of a pulsating jet in the external wave field in the vicinity of the open end, and so on. Some investigations involving aerosols in large volumes were performed for conditions of intense harmonic excitation outside of the resonance frequencies of oscillation in work [7]. Most interesting is the case of nonlinear resonance oscillation of aerosols in tubes; however, only several studies were devoted to experimental study of this phenomenon [16-27]. The behavior of the aerosol with its nonlinear oscillations in a closed tube near the first eigenfrequency was studied in experimental works [16-18, 20-26]. Some results are given in [19, 21, 22, 27] of an experimental investigation of oscillations in closed and open tubes near the subharmonic resonance, the frequency of which is two times lower than the first eigenfrequency. The main nonlinear effect was accelerated the coagulation and the deposition of the machine oil droplets and tobacco smoke [16], oleic acid droplets [17], smoke particles produced by the combustion of an incense stick [18], and droplets obtained from liquid di-ethyl-hexyl-sebacate (DEHS) [19-27] on the tube walls. The geometric diameter of the aerosol droplets and particles was 1–10 μm [16, 17], 0.3 μm [18], and 0.863 μm [19-23].

This work is aimed at experimental study of the features of coagulation and deposition of the oscillating aerosol near the first eigenfrequency in the closed and the open tubes in the shock-free mode.

2. Experimental conditions

The longitudinal formed oscillations of an aerosol in a vertical quartz tube of length L = 1.06 m and of internal diameter $2R_0$ = 0.0365 m are created by a planar piston. It is put into motion on a vibrostand with a power amplifier TV51075 (TIRA). The passive end of the tube is covered by the cap for studies in a closed tube. The cap with an orifice, coinciding with diameter of the tube for studies in an open tube is used. The measurements and control of the sinusoidal oscillations are performed in real time on the basis of software SineVIEW, by means of a piezoelectric IEPE accelerometer 4513-001 (Bruel & Kjaer) and a controller VR8500-1 (Vibration Research Corporation). The values of the frequency ν and amplitude of the piston position l are set with an accuracy of better than 10^{-6} Hz and 10^{-7} m, respectively. The pressure measurement system consists of a pressure sensor 8530C-15 (Bruel & Kjaer) with a diameter of 3 mm placed near the piston, a three-channel bridge voltage amplifier ENDEVCO-136 (Bruel & Kjaer), and a digital oscilloscope DSO 3062A (Agilent Technologies). Oscilloscope control and recording of the shape and values of the output signal with an accuracy of better than 0.3 mV is performed also by a computer, using the software DSO3000. To determine the transparency of the aerosol a measuring complex is installed, in which a laser with the wavelength of 630 nm and power 0.16 W with a remote control is used as a light source and a digital luxmeter ATT-1505 (Aktakom) is used as a receiver. Di-ethyl-hexyl-sebacate C26H50O4 is employed as a working fluid. The aerosol is created by an aerosol generator ATM 225 (TOPAS). The aerosol droplets size was in range from 0.839 to 1.06 μm. The majority of droplets have the geometric diameter D = 0.863 μm.

The measurements are performed for different values of the amplitude of piston travel in the frequency range, containing resonance. The measured resonant frequencies ν^* = 160.7 Hz for the closed tube and ν^*= 78.6 Hz for the open tube due to the near-wall losses are different from the values ν_1 = 161.8 Hz and ν_1 = 80.1 Hz, calculated on the basis of the linear acoustics.

The experiments are performed in two stages. At the first stage, the parameters of the pressure oscillations are measured. The output signal from the pressure sensor is recorded by given values of the frequency and of the amplitude of the piston oscillations. At the second stage, the transparency of an oscillating aerosol is measured. The light source and the light sensor of luxmeter are mounted on the racks so that the light beam passed through the medium across the tube and fell into the center of

Journal of Physics: Conference Series **567** (2014) 012021 doi:10.1088/1742-6596/567/1/012021

the sensor. As an initial value of the luxmeter signal Φ_0 the signal by the illumination of the tube, filled with an air, is used. The tube is filled with the aerosol by means of the aerosol generator, without applying an additional pressure. Duration of load the tube with the aerosol is chosen so that the initial value of the concentration was equal to 3.5×10^6 cm^{-3} in all experiments. The luxmeter signal Φ_1 is recorded during the oscillations. The change of the aerosol transparency occurs in the experiments, which causes the change of the luxmeter signal Φ_t. It was assumed that the correlation between the illuminance and the numerical concentration of aerosol droplets in the tube is linear and, therefore, the relation $N = 3.5\times10^6\ (\Phi_t - \Phi_0)/(\Phi_1 - \Phi_0)$ cm^{-3} can be used for the calculation of the numerical concentration of the aerosol droplets.

3. Experimental results

The numerical concentration of the droplets monotonically decreases with time. And under natural deposition a slower decrease in the aerosol concentration with time takes place. The obtained dates are used to determine the time scale of the coagulation and the deposition of aerosol for the closed tube and the time scale of the discharging of aerosol for the open tube as the functions of the frequency oscillations and amplitude of piston travel.

The dependences of the time scale of the coagulation and the deposition of aerosol and the time scale of the discharging of aerosol on the parameter ε at the frequencies of the observed resonance are given in Figure 1. It is seen that the increase in ε leads to nonlinear reduction of the time scale of the coagulation and the deposition of aerosol in the closed tube (Figure 1(a)) and almost linear decrease in the time scale of the discharging of aerosol in the open tube (Figure 1(b)). It is possible to note that the time scale of the coagulation and the deposition of aerosol is 2-4 times, and the time scale of the discharging of aerosol is 6-12 times less than time scale of the natural deposition of aerosol in the absence of oscillations. Effective coagulation of aerosol droplets and their deposition in the tube due to the secondary flow in the manner of two toroid vortices [15], formed as a result of the nonlinearity of the resonant oscillations, is observed. This can be explained so that the medium oscillation corresponds to the harmonic law at the small amplitude of piston travel in the closed and in the open tubes: the shape of the pressure wave is continuous and symmetrical. The amplitude of the pressure oscillations increases with increasing amplitude of piston travel. In the closed tube, the wave shape is distorted and in the open tube, the shape of the pressure wave takes the form somewhat different from harmonic.

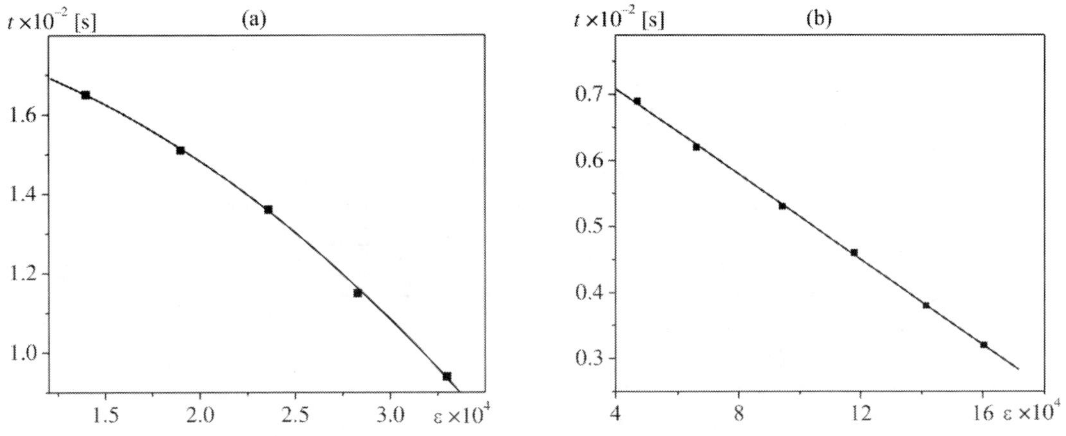

Figure 1. Dependence of the time of coagulation and deposition of aerosol (a) and the time of discharging of aerosol (b) on the parameter ε at observed resonant frequency: the line is the polynomial approximation.

Figure 2 gives dependence of time scale of the coagulation and the deposition and the time scale of the discharging of aerosol on the excitation frequency for the amplitude of piston travel l = 0.3 mm, where each point corresponds to a separate experiment. The dependence demonstrates a nonmonotonic behavior with a minimum value at resonant frequency. This is because the amplitude of aerosol oscillation increases close to the resonance. This dependence is explained by the fact that close to the resonance the amplitude of the forced oscillations of aerosol increases, which reduces the process time. At the resonance, the amplitude reaches the maximum value, and the process time has a minimum value. Over the resonant frequency the amplitude decreases again, resulting in an increase in the process time. It should be noted that the time scale of discharging of aerosol droplets in the open tube about 1.5 times less than the time scale of coagulation and deposition of aerosol droplets in the closed tube. This results in the fact that in the open tube the aerosol is released into the environment.

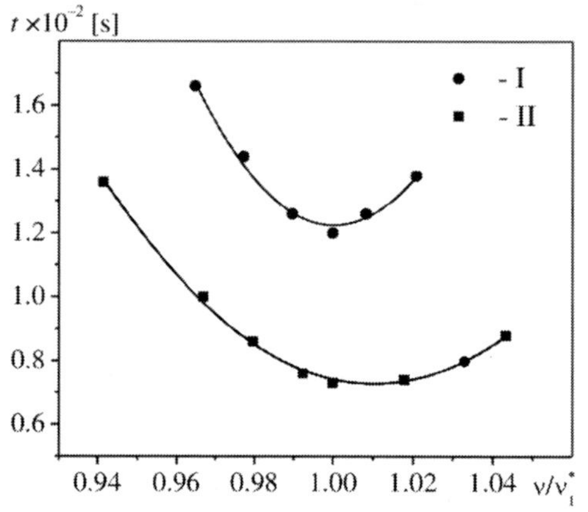

Figure 2. Dependence of time scale of the coagulation and the deposition (I) and time scale of the discharging (II) of the aerosol on the excitation frequency for the amplitude of piston travel l = 0.3 mm: the line is the polynomial approximation.

4. Conclusions

It is shown that the presence of oscillations leads to acceleration of the coagulation and the deposition of aerosol. The increase in intensity of the oscillations, caused by increase in amplitude of piston travel leads to reduction of process time. It is found that time scale of coagulation and deposition of the aerosol droplets in the closed tube is 2-4 times and the time scale of discharging of aerosol in the open tube is 6-12 times lower than by natural deposition. In so doing, the time scale of the discharging of aerosol in the case of the open tube is reduced by a factor of 1.5 and more compared to the time scale of the coagulation and the deposition in a closed tube. The nonmonotonic behavior of the dependence of the time scale of the coagulation and the deposition of aerosol in the closed tube and the time scale of the discharging of aerosol in the open tube on the frequency of excitation with a minimal value at the resonant frequencies is observed.

Acknowledgments
This work was funded by the subsidy of the Russian Government to support the Program of Competitive Growth of Kazan Federal University among World's Leading Academic Centers.

References

[1] Nigmatulin R I 1990 *Dynamics of Multiphase Media* (New York: Hemisphere)
[2] Ganie R F and Ukrainskii L E 1975 *Dynamics of Particles under the Effect of Vibration* (Kiev: Naukova Dumka)
[3] Ganiev R F, Kobasko N I et al. 1980 *Oscillation Phenomena in Multiphase Media and Their Uses in Technology* (Kiev: Tekhnika)
[4] Gubaidullin D A 1998 *Dynamics of Two-Phase Gas-Vapor-Droplet Media* (Kazan: Kazan Mathematical Society)
[5] Gubaidullin D A and Nigmatulin R I 2000 *Int. J. Multiphase Flow* **26** 207
[6] Bergmann L 1938 *Ultrasonics* (London: Bell)
[7] Mednikov E P 1965 *Acoustic Coagulation and Precipitation of Aerosols* (New York: Consultant Bureau)
[8] Fuks N A 1955 *Aerosol Mechanics* (Moscow: USSR Acad. Sci.)
[9] Voloshchuk V M and Sedunov Yu S 1975 *Coagulation Processes in a Dispersed System* (Leningrad: Gidrometeoizdat)
[10] Voloshchuk V M 1984 *Kinetic Theory of Coagulation* (Leningrad: Gidrometeoizdat)
[11] Temkin S 2005 *Suspension Acoustics: an Introduction to the Physics of Suspensions* (New York: Cambridge University Press)
[12] Shaw D T 1978 *Acoustic Agglomeration of Aerosols, in Recent Developments in Aerosol Sciences* (New York: Wiley Interscience)
[13] Hoffmann T L and Koopmann G H 1996 *JASA* **99** 2130
[14] Hoffmann T L and Koopmann G H 1997 *JASA* **101** 3421
[15] Ilgamov M A, Zaripov R G, Galiullin R G and Repin V B 1996 *Appl. Mech. Rev.* **49** 137
[16] Gulyaev A I and Kuznetsov V M 1963 *Sov. Phys. Acoust.* **8** 236
[17] Temkin S 1970 *Phys. Fluids* **13** 1639
[18] Shuster K and Fichman M, Goldshtein A and Gutfinger C 2002 *Phys. Fluids* **14** 1802
[19] Gubaidullin D A, Zaripov R G, Galiullin R G, Galiullina E R and Tkachenko L A 2004 *High Temperature* **42** 794
[20] Gubaidullin D A, Zaripov R G and Tkachenko L A 2012 *High Temperature* **50** 564
[21] Gubaidullin D A, Zaripov R G and Tkachenko L A 2013 *High Temperature* **51** 955
[22] Gubaidullin D A, Zaripov R G and Tkachenko L A 2013 *Doklady Physics* **58** 392
[23] Gubaidullin D A, Zaripov R G and Tkachenko L A 2012 *J. of Engin. Physics and Thermophysics* **85** 259

Research on influence of temporary distribution of power of laser radiation with a hybrid processing on a depth heat affected zone.

D A Bashmakov, I H Israphilov, A T Galiakbarov
Department of Energy and Information, Naberezhnochelninsky Institute (branch) Kazan (Volga Region) Federal University, pr. Mira 68/19, Naberezhnye Chelny, 423810, Russian Federation

E-mail: DABashmakov@kpfu.ru

Abstract. Investigations on the influence of the temporal distribution of the laser radiation power per pulse of hybrid processing on a depth heat affected zone. Recommendations on use of specific distribution of the laser radiation power in the combined laser and plasma processing for decrease in energy consumption.

1. Introduction

One of the most effective methods of treating metal alloys is to use hybrid processing methods using plasma and laser radiation flux [1]. The main criterion for assessing the quality of the process parameters of the hybrid hardening serves temperature distribution in the interaction zone of concentrated energy flows, and from it has, in turn, depend all the other indicators of the quality of the hardening [2, 3].

In practice there is often a question of depth heat affected zone (HAZ), particularly for machine parts operating on an abrasion. Surface treatment of parts using hybrid laser-plasma method in the critical energy density of the laser beam, do not give stable quality indicators, which leads to the inability to use these technologies in the industry.

For ensuring high-quality management of technological processing of a product, it is necessary to study more fully the processes proceeding at influence of laser radiation [4]. In this work research of influence of temporary distribution of power of laser radiation to depth of a zone of phase transformations in metal at influence of the laser is conducted.

2. Experimental study

For carrying out experiments with a research objective of interaction of laser radiation, with various temporary distribution of power, with metals experimental installation on the basis of the laser technological complex LRS-150A was designed (figure 1).

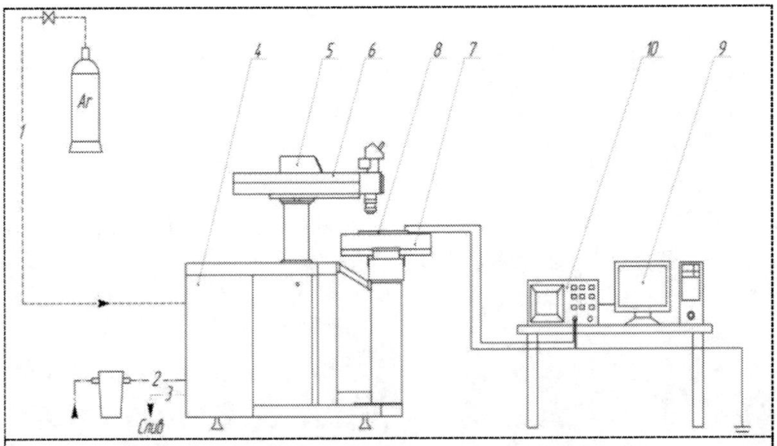

Figure 1. Experimental laser technological machine.

In figure 1: 1 - Argon supply; 2 - Supply of cooling liquid; 3 - Water plums; 4 - Power unit and coolings of the laser; 5 - Control panel; 6 - A laser radiator with the optical control focusing system; 7 - The automated two-coordinate table; 8 - Target sample steel 45; 9 - The operating computer; 10 - Digital oscillograph of ASK-2034.

The used laser technological complex allows to generate pulse laser radiation with various temporary distribution of power in an impulse. For carrying out experiment were used the most widespread and encountered in using in the industry of distribution of power of radiation in time: uniformly, Gaussian, evenly increasing and evenly decreasing. To achieve the correct measurements, energy input to the pulse was the same, which is achieved by the same voltage capacitor charging power supply system of the laser technological complex.

For completeness of consideration of the processes proceeding during radiation on a detail were two operating modes which are characterized various voltage of a charge condensers U_K and time of pulse influence t are picked up: U_K = 420 V, t = 4 ms и U_K = 300 V, t = 2,5 ms. Steel 45 microstructure, when processing with use of the modes described above, are presented in figures 2 and 3.

3. Results of experiment

In figures 2 and 3: a – uniform distribution, b – Gaussian distribution, c – the increasing distribution, d – the decreasing distribution.

The analysis of the processed samples at U = 300 V, t = 2,5 ms showed to the following dependencies which are possible to distinguishing: at the uniform and decreasing temporary distribution of power in an impulse of laser radiation almost identical depth of the zone of thermal influence.

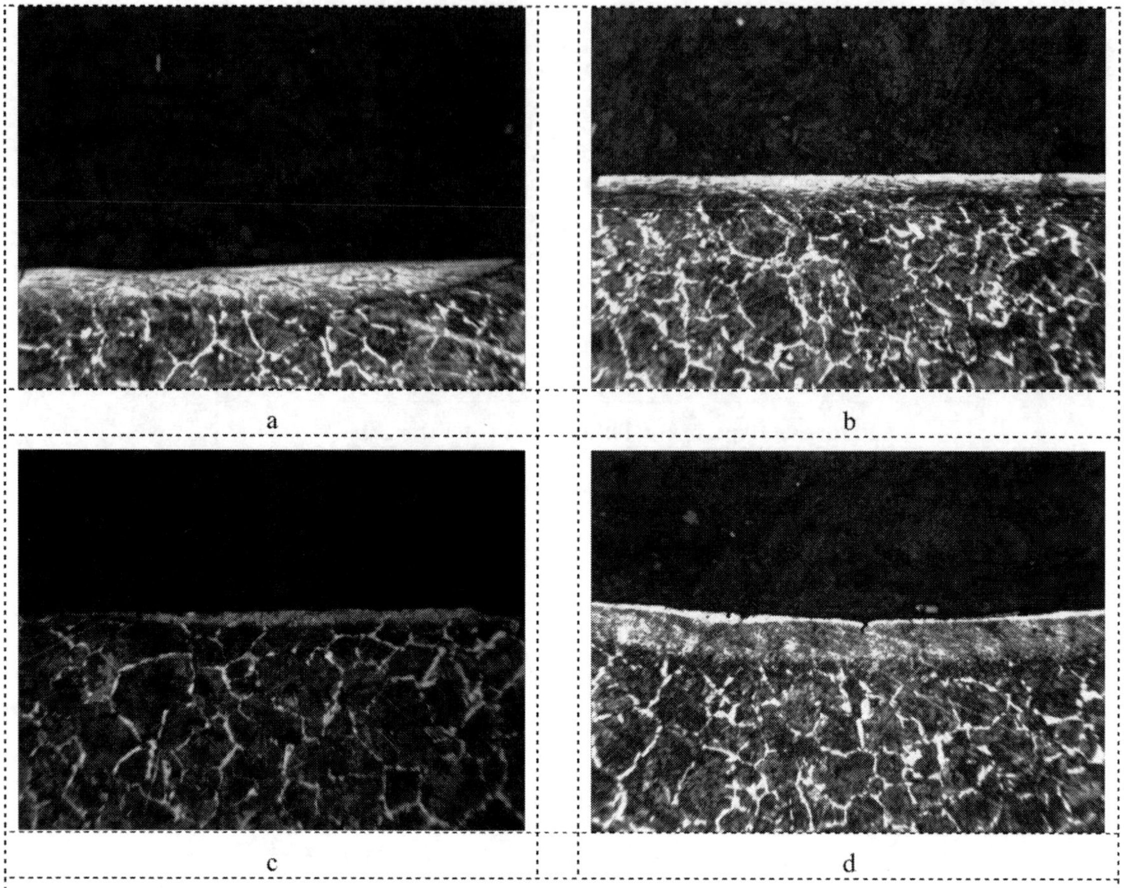

Figure 2. Microstructure of a zone of laser influence $U_K = 300$ V, t = 2,5 ms, x200

Under the influence of the Gaussian and the increasing temporary distribution of energy, depth of HAZ is much less, in difference from the uniform and decreasing distribution.

The values obtained for the depth of the HAZ when $U_K = 420$ V, t = 4 ms (Table 1) are similar in nature to the results of the impact when $U_K = 300$ V, t = 2,5 ms, but the difference in the depth of the HAZ impact at different temporal distribution of power laser pulse is reduced.

We will note that the samples processed with the decreasing temporary distribution of energy have the greatest depth of HAZ, and the samples processed with the increasing temporary distribution have the smallest depth of HAZ.

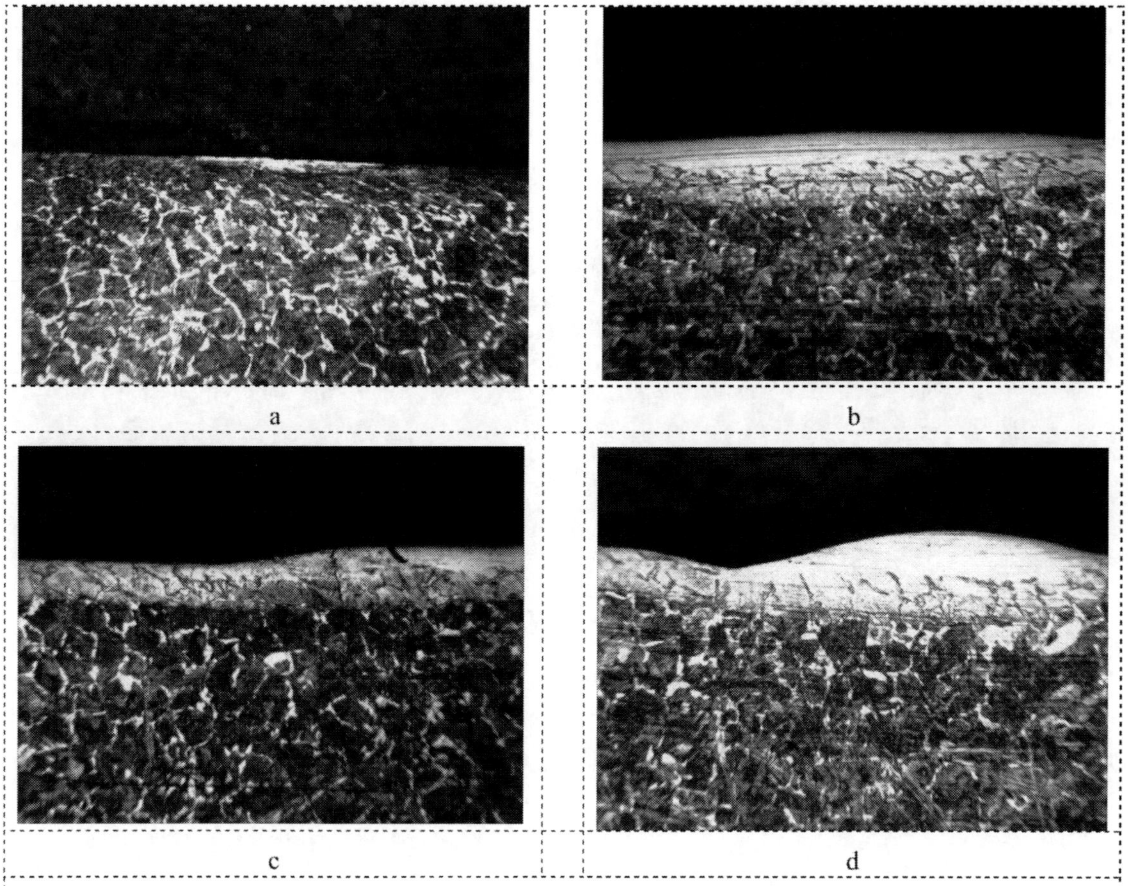

Figure 3. Microstructure of a zone of laser influence U_K = 420 V, t = 4 ms, x200

Table 1. The dependence of the depth of HAZ of temporary power distribution in the laser pulse.

Temporary power distribution	Depth of HAZ, h, μm	
	U_K = 300 V, t = 2.5 ms	U_K = 300 V, t = 2.5 ms
Uniformly	65.4	18.7
Gaussian	37.4	89.7
Evenly increasing	22.4	74.8
Evenly decreasing	71	97.2

4. Conclusions

Thus, at hybrid laser and plasma processing of materials, for decrease in energy consumption of technological installation and management of stability indicators of quality processing, it is necessary to strive for the absence of increasing laser power during the whole cycle of continuous exposure to laser radiation. In case of pulse laser influence in the combined power source, it is necessary strives for the falling characteristics of power in an impulse of laser radiation.

References

[1] Israphilov I H, Bashmakov D A, Galiakbarov A T, Mandrik P A and Ganiev M M 2014 *Russian Physics Journal* **57** **3/3** 148
[2] Grigor'yants A G, Shiganov I M and Misyurov A I 2006 *Technological processes of laser processing* (Moscow: Bauman Moscow State Technical University) p 664
[3] Israphilov I H, Galiakbarov A T, Israphilov D I, Gabdrahmanov A T and Samigullin A D 2012 *Proceedings of the Tula State University. Technical sciences.* **6** 90
[4] Israphilov D I 2007 *Management of a plasma technological complex for heat treatment of parts with specified strength characteristics* (Naberezhnye Chelny: Kama state engineering and economic academy) p 141

The spatial distribution the thickness of polymer powder coatings for ultrasonic sensors

V A Gavrilova, M G Fazlyyyakhmatov and N F Kashapov
Kazan (Volga Region) Federal University, 18 Kremljovskaja str., Kazan 420008, Russian Federation

E-mail: mfazlyjy@kpfu.ru

Abstract. Objects of research are coatings and technology of their applying to the piezoelectric elements for ultrasound. Results of studies the distribution coating thickness according to different modes of coating process are presented. Experimentally confirmed the simulation results of the movement gas suspension on the electrostatic field in the electrode system "needle - plane".

1. Introduction

Ultrasound remains one of the most popular tools non-destructive testing and diagnostics in various areas including medicine. Our research group conducts researches in the field of ultrasound and the development of new methods and equipment for non-destructive testing of multiphase and biological objects [1-5].

The source of ultrasonic vibrations are electromechanical transducers based on piezoelectric materials. Despite the appearance of new piezoelectric composite materials and various polymer piezoelectric films the most common materials remains barium titanate ($BaTiO_3$) and lead zirconate titanate (PZT). The main problem of these materials is the brittleness.

Close contact ultrasonic sensor with the test object is necessary in the process of diagnosis liquid and solid objects. This requires to use different protective coatings and prisms which prevent the damage transducer and protect it from aggressive environments. For example, in the medicine copolymers of vinyl chloride and ethyl acrylate are applied. Plates of which are glued to the piezoelement. However, the protective coatings are abraded and are damaged, resulting in a failure of the sensors. To solve this problem a corona discharge plasma for drawing polymer powder coating has been proposed [6]. Advantages of the method of plasma spraying in an electrostatic field are ease the adjustment process, the possibility of automation, high productivity, lack of pre-heating products, uniform coating thickness, the ability to produce coatings on thin-walled products, as well as products made from different materials [7].

2. The protective layer thickness calculation

The important function of the coating applied to the piezoelectric element is the matching of the acoustic impedances of the piezoelectric element and the test environment. It is necessary to reduce the loss upon reflection at the interface. This matching is carried out by introducing "quarter wave" layer. Waves which reflected from both boundaries matching layer are in antiphase and will have the same amplitude. This will lead to their mutual compensation. The phase shift by 180 degrees is provided the path difference of the waves between the boundaries in half of wavelength, and the equality of the amplitudes - rational choice acoustic impedance transforming layer [8].

For most tasks of ultrasound diagnostics and probing the frequency range 0.5-15 MHz is used. Ultrasonic vibrations in this frequency range provide sufficient resolution in the low level vibration damping.

The average propagation velocity of ultrasonic vibrations in human tissues is 1540 m/s, in water (at t = 20^0 C) 1480 m/s, in transformer oil is about 1390 m/s. The transformer oil is used for the simulation experiments, as the location of the piezoelectric elements in the water leads to disruption of their acoustic properties. The expression for calculating the acoustic impedance matching layer:

$$Z_{tr} = \sqrt{Z_1 Z_2}\,, \tag{1}$$

where Z_1 – the acoustic impedance of the first medium, Z_2 – the acoustic impedance of the second medium.

For water, the acoustic impedance $Z = 1{,}483 \times 10^6\ kg/m^2\,\mathrm{s}$, for piezoelectric ceramics PZT-19 $Z = 22{,}42 \times 10^6\ kg/m^2\,\mathrm{s}$, respectively, the acoustic impedance the polymer powder coating should be about $Z_{tr} = 5{,}77 \times 10^6\ kg/m^2\,\mathrm{s}$.

3. Preparation of the samples and selection of spraying system parameters

In this paper we investigated the polyether paint coating PD810119G («BECKER»). This paint corresponds safety requirements due to the absence of organic solvents and heavy metals.

Powder paint - is the solid dispersed multicomponent system, which includes a special film-forming resins, hardeners, pigments, fillers and special purpose additives. For medical products polyether, epoxy-polyester coatings is used. They have the following characteristics: resistance to abrasion, impact resistance, smoothness and convenience the surface, non-toxicity, protection from the bactericidal activity, suitability for sterilization, ease for cleaning, chemical resistance.

Getting coatings requires careful surface preparation. The process of preparing the surface discussed in [10]. Surface must be clean, dry and free of contaminants.

Spraying the coating is carried out using a spray gun. The spray gun - the unit for applying the polymer powder paint that provides to obtain a powder mixture with air, the formation of the torch and giving the powder particles an electric charge.

Many properties of polymer powder coatings are largely dependent from the dispersion of the powder paint. As a method of determining dispersion microscopic analysis is widely used. The main advantage is that it allows to determine not only the geometrical dimensions of the studied objects, but also to see the features of their shape, surface structure.

Research and microscopic analysis of the particulate composition of the powder paint for electrostatic spraying showed that the most suitable powder coating for the application on piezo element ultrasonic sensors is such that the particulate composition which is in the range of 12-35 microns [11]. Was found that the powder paint particles the size of 1-12 microns during the spraying carried away on the air filter unit of the spray chamber due to the small weight. And the powder paint particles larger than 35 microns are showered from sprayed surface under gravity.

The studies [7] of physical and mechanical properties the samples of the coatings of polyether paint PD510226 («BECKER»), epoxy-polyether paint EP110022G («BECKER») showed the suitability of using this type of coating to the surface of the ultrasonic transducers. Such parameters as impact strength, flexibility, adhesion, gloss, hardness and wear resistance were evaluated

Spraying is carried out in the chamber of coating polymer powder materials at the samples. Spraying occurs due to sticking of the electrified polymer particles with the sample. The subsequent formation of the coating is carried out by heating.

The polymerization process occurs in the drying chamber, where the product is heated, fusion the polymer particles and the final formation of the coating occurs.

Journal of Physics: Conference Series **567** (2014) 012023 doi:10.1088/1742-6596/567/1/012023

The measurement of the potential electrostatic field during application polymer powder coatings on metallic substrates using different speeds of micro-particles and the distance to the substrate have been carried out [12]. The current-voltage characteristics for different flow speeds in the discharge, as well as without spraying at various distances from the substrate are obtained. The current-voltage characteristic of corona discharge during application of polymer powder coatings shows that the corona discharge ignition occurs at a voltage of 5 kV (the value of the discharge current is not higher than 0.5 mA) at a voltage of 5 - 10 kV, the discharge current has a value not exceeding 5 mA at various flow rates; in the range of 10V - 30 kV is visible smooth increase of the discharge current at a flow rate of 8 - 22 m/s; maximum discharge current of 25 mA is achieved at a voltage of 30 kV and a flow rate of 22 m/s.

The concentration of solid particles close to processed surface depends on its shape, the distance and potential difference between surface and electrode and the spray gun, the concentration of the carrier and the dispersed phase in the flow, dimensions, density and charge aerosol particles.

Modelling of movement the gas suspension on the electrostatic field in the electrode system "needle - plane" is produced [13]. As a model that takes into account the above factors was chosen system of equations collisionless dynamics of monodisperse two-temperature two-speed medium without phase transitions.

Is shown that the influence of electrostatic forces on the flow of gas suspension is observed only in close proximity to the electrodes where the potential gradients are greatest. In the interelectrode space in these charge density of the particles dominate the aerodynamic forces. The presence of the Coulomb force near the surface can reasonably set on the surface adhesion conditions for the velocity of the particles.

Except the above parameters protective matching layer of ultrasonic sensors, the coating thickness is also important. It must correspond to a quarter of the wavelength emitted and received by the sensor. The performance characteristics coating system were established and the conditions of their formation are proposed [8].

4. Experimental results and discussion

The experiments to measure the spatial distribution of coating thickness produced in the different modes are performed (Figure 1a, 1b and Figure 2). For this the prepared metal substrate (size 7x10 mm) was coated by PD810119G polyether paint. Polymerization was carried out in the drying chamber at 200 ° C within 10 min.

The measurement of thickness the obtained coatings was carried out symmetrically relative to the centre of the substrate to the left and to the right in increments of 5 mm using an ultrasonic thickness gauge PosiTector 200 B/Adv. This device allows to measure the thickness of the coatings in the range from 13 to 1000 microns with an accuracy ± (2 microns + 3% from the measurement value).

Figure 1 shows that the coating is heterogeneous with a maximum thickness at the centre, followed by a decrease and an increase closer to the substrate edge.

Conducted calculations [13] showed that during application the aerosol to a flat surface oriented normal to the axis spraying in an electrostatic field, the solid phase is applied nonuniformly and its density increases towards the free edge. At free edge the solid phase reaches the highest value. This is defined bigger intensity of the electrostatic field in this area. The experimental results confirm these calculations.

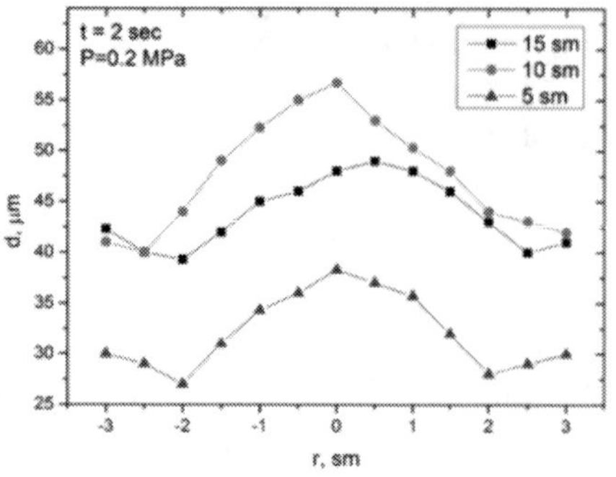

Figure 1a. The distribution of the coating thickness at different distances spray gun to the substrate: spraying time t=2 sec, pressure on the gun P=0.2 MPa

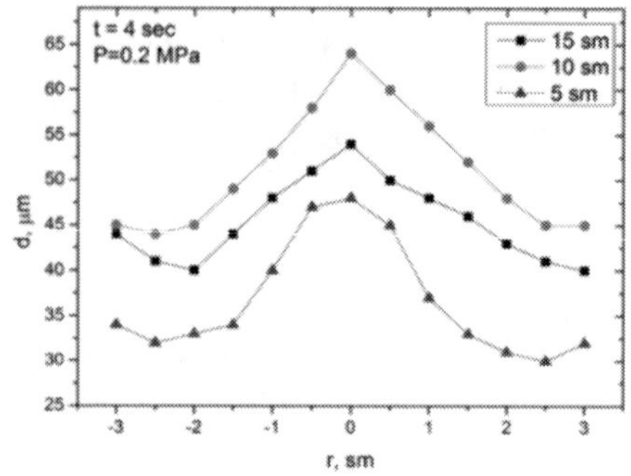

Figure 1b. The distribution of the coating thickness at different distances spray gun to the substrate: spraying time t=4 sec, pressure on the gun P=0.2 MPa

In the absence of a field, the greatest density of the dispersed phase is reached in the internal surface spraying area that is connected to the configuration of the vortex flow formed in front plate [13]. This is most likely explained by the fact that at the centre of the substrate exhibits a maximum thickness, i.e. the particles deposited screen the electric field, and the particles are the next depositing according to the configuration the vortex flow.

In both modes, the maximum coating thickness is obtained at a distance from the needle the spray gun to the applied surface equal to 10 cm. With further increase distance the thickness of the coating sharply decreases, and measure it used devices failed. With decreasing distance to 5 cm thickness of the coating is also strongly decreases and becomes highly inhomogeneous. This is because the aerodynamic forces the flow of the gas suspension impede of paint particles sticking to the surface blowing them.

With increasing deposition time (Figure 2) it becomes possible to increase the distance from the needle the spray gun to the applied surface.

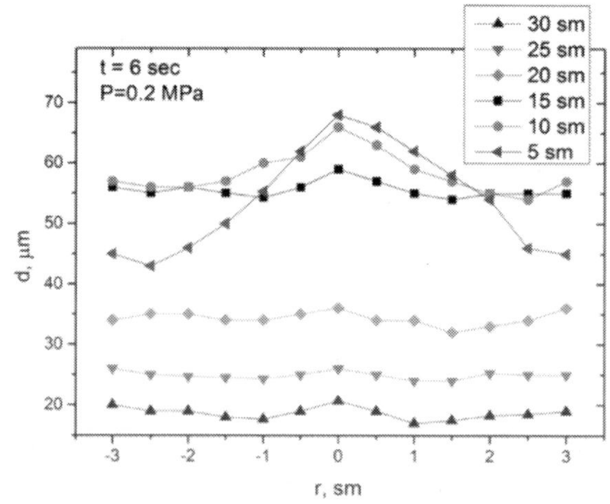

Figure 2. The distribution of the coating thickness at different distances spray gun to the substrate: spraying time t=6 sec, pressure on the gun P=0.2 MPa

Figure 2 shows that increasing the spraying time to 6 seconds leads to an increase in the thickness of the coating and increase its homogeneity. This is because during prolonged spraying occurs shielding the electric field on the surface of the substrate settled by particles of paint.

Simulation results [13], and the experimental data shows that the coating is heterogeneous by thickness, which is rather critical for solving our problem. However, it is possible to select and use the various modes of spraying, depending from the geometry of the ultrasonic sensor.

4. Conclusions

Results of studies the distribution coating thickness according to different modes of coating process are presented. Experimentally confirmed the simulation results of the movement gas suspension on the electrostatic field in the electrode system "needle - plane". The modes to obtain polymer powder coating of ultrasonic sensors for different geometric dimensions is proposed.

References

[1] Denisov E S, Sagdiev R K, Evdokimov Yu K, Temyanov B K and Fazlyyyakhmatov M G 2013 *Proc. Int. Sci. and Tech. Conf. Nigmatullin's Readings -2013* (Kazan: Kaz. St. Tech. Un. Press) 134-137

[2] Fazlyyyakhmatov M G and Kashapov N F. 2011 *Proc. III Sci. and Tech. Conf. Low Temperature Plasma in Pr. Functional Coatings* (Kazan: Kaz. Nat. Res. Un. Press) 97-101

[3] Fazlyyyakhmatov M G and Kashapov N F. 2014 *Proc. Int. Sci. and Tech. Conf. 50 Year of MRTI-BSUIR-2014* (Minsk: BSUIR Press) **2** 119-120

[4] Fazlyyyakhmatov M G and Kashapov N F. 2014 *Proc. 18 Int. Young Forum Radioel. And You In 21 Cent.* (Kharkov: KhNURE Press) **1** 124-125

[5] Fadeev S.A. and Kashapov N.F. 2013 *J. Phys.: Conf. Series* **479** 012010

[6] Gavrilova V A, Kashapov N F 2010 *Vestnik KGTU* **7** 117-125

[7] Gavrilova V A, Kashapov N F, Kashapov R N 2011 *Biomedical Engineering* **45** 198-200

[8] Gavrilova V A, Fazlyyyakhmatov M G and Kashapov N F 2013 *J. Phys.: Conf. Series* **479** 012010

[9] Yakovlev A D 1987 *Powder paints* (Leningrad: "Chemistry") p 216

[10] Gavrilova V A, Fazlyyyakhmatov M G and Kashapov N F 2013 *Repair, Rebuild, Modernization* **12** 13-16

[11] Gavrilova V A, Kashapov N F and Volgaev S A 2011 *Russian Coatings Journal* **8** 38-42

[12] Gavrilova V A, Fazlyyyakhmatov M G and Kashapov N F 2014 *Russian Physics Journal* **57** **(3/3)** 119-123

[13] Gavrilova V A, Kashapov N F, Tukmakov A L and Tukmakov D A 2012 *Vestnik KGTU* **2** 177-182

Influence of plasma-electrolyte discharge to the glass surface

L N Kashapov[1], N F Kashapov[1] and R N Kashapov[1,2]
[1] Engineering Institute , Kazan Federal University, Kazan, 420008, Russia
[2] Laboratory of Radiation Physics, Kazan Physical-Technical Institute of the Kazan Scientific Center of the Russian Academy of Sciences, Kazan, 420029, Russia

E-mail: kashramiln@gmail.com

Abstract. Gas discharges with liquid electrodes are known more than a hundred years. For these discharges typical is the fact that one of the electrodes is a conductive solution of electrolyte. The research resulted in found that the mechanisms of removal of glass for anodic and cathodic regimes differ. At cathodic regime occurs mainly thermal effect from the discharge. And at anodic regime occurs mainly glass etching by plasma of discharge.

1. Introduction

Gas discharges with liquid electrodes are known more than a hundred years. For these discharges typical is the fact that one of the electrodes is a conductive solution of electrolyte. To initiate the discharge electrode (at which combustion will take place) and the "counter-electrode" (larger area than first) omitted in the electrolyte solution. Further a voltage is applied to them and occurs an ignition of the discharge. This type of discharges has been widely used in various branches of science and engineering. One of the branches of the active application is the processing of metals and their alloys, as well as infliction of functional coating. Treatment of non-conductive materials was considered, in principle, not feasible. This problem was partially solved in 1968, by team of Japanese scientists [1]. This treatment has been called chemical-spark engraving. In recent years this theme has been actively developed by Canadian scientists [2, 3, 4]. In all of these studies research conducted from the viewpoint of the effects of thermal flux which generated by the discharge in the treatment area. It does not investigate the influence on parameters of discharge burning on treatment process and have not been studied parameters of low-temperature plasma.

Therefore the purpose of this work is to study the parameters of the discharge in the process of plasma-electrolytic engraving glass.

2. Experimental

To achieve this purpose has been developed the experimental setup on the basis of 3D-printer Picaso Builder. Schematic diagram of the setup is shown in Figure 1.

To control the coordination system of the 3D printer was used program Polygon. This system was chosen based on the simplicity of design and handling. Also in the event of failure of any node, it is easily replaced. Instead of the standard print head was manufactured device for mounting the electrode on which the generations of plasma-electrolyte discharge (Fig. 1B). Processing was conducted in an electrolytic bath (h=30 mm, l=215 mm, d=215 mm) Bath fixed on the surface of the platform 3D-printer. In the bath were located dielectric pillars on which processed glass are fixed. Also in the bath been two holes for pumping the electrolyte.

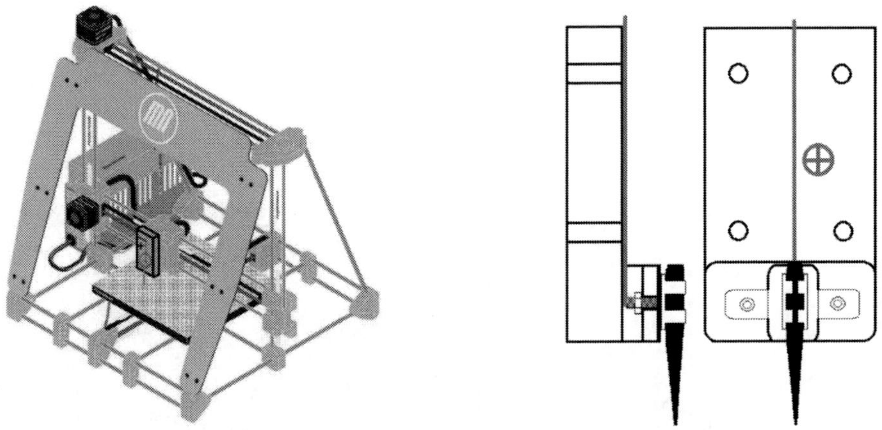

A B

Fig.1. The experimental setup of engraving glass (A), the head of the generation of plasma-electrolyte discharge (B).

One at the bottom and the second at the top. Thus, the hot electrolyte used in plasma-electrolytic process is gradually replaced by a new cold. For pumping of the electrolyte used a peristaltic pump. On the two sides of the electrolytic cell installed copper counter-electrode with dimensions 20×100×1,5 mm. As a electrode, at which the occurs initiation of plasma-electrolyte discharge, was used cylinder made of steel 12X18H10T sharpened by a cone with an angle of 3.5 degrees and a radius of the top 30 microns. Electrolytic bath filled by researched electrolytes with necessary concentration and composition. The working fluid used a 5% aqueous solution of alkali NaOH.

The whole complex is consisted of a system of electric power, the electrode system, an oscilloscope, a shunt, the electrolytic bath, thermometer and two digital multimeters.

The system power supply is a DC power source with continuously variable voltage. The power supply consists of the diode bridge and laboratory autotransformer adjustment type 1M with a voltage range from 1 to 240 V, 50 Hz frequency. The main operating parameters of the electrode system: the voltage across the electrodes, the magnitude of the discharge current, the current density at the electrodes.

Used a pulsating current source obtained after full-wave rectification. Changing the form of voltage and current at the time of discharge ignition was determined by the oscilloscope FLUKE 105 SCOPEMETER SERIES II, time base varied from 5 ns to 60 s. The relative measurement error did not exceed 0.025%.

Measurement of the voltage at the anode is made through digital multimeter APPA 207, the relative measurement error is 0.2%. With a DMM APPA 305 we measured current of plasma-electrolyte discharge. The temperature distribution near the surface of the active electrode was measured by an infrared thermometer Fluke 62 Max. The principle of operation in the infrared pyrometer is based on measuring the thermal energy of objects, its main body is a sensor which detects the radiation from a heated body in the infrared range and converts it into an electrical signal proportional to temperature. Further this signal is processed, the result is displayed on the built-in display. Treatment of the obtained results was performed taking into account the measurement uncertainty.

3. Results

Figure 2 shows the current-voltage characteristic of the plasma-electrolytic process without contact with the glass and in contact with the glass.

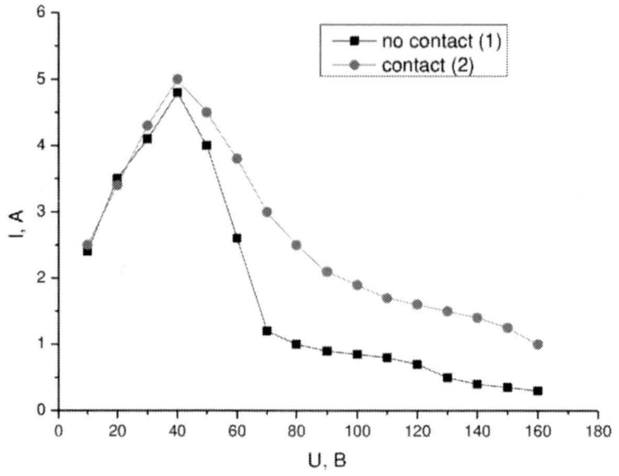

Fig.2. CVC of plasma-electrolytic process with a smoothing capacitor filter. 1 - without contact with the glass, 2 - the electrode is in contact with the glass

In this case, the discharge initiation performed on the metal anode, when the cathode was the electrolyte. Consider a more detailed view of the current-voltage characteristics. They have a maximum which characterizes the phenomenon of transition from the electrolysis process to the formation of steam-shell and discharge burning. In considering the CVC process without contact with the glass surface shows that discharge burning in the vapor-sheath occurs at a voltage of 70 V and a current of 1.1 A and a further increase in voltage leads to a decrease in current. Reducing the current associated with a decrease in amount of of charge carriers, and it may be due to increasing thickness of the steam-shell (the length of the electrode gap). The second curve describes CVC of plasma-electrolytic process, when the tip of the metal anode in contact with the surface of the glass. The shape of the curve remains the same, but the steam-shell formation and initiation of discharge burning was observed at higher values of voltage and current. This CVC coincides with the curve of CVC of plasma-electrolytic process by using full-wave rectification mode without smoothing capacitor filters [5]. This coincidence suggests about the same effect on formation of steam-shell electrode shape and form of the applied voltage. In [6], studies have been conducted of the transition from the regime of electrolysis to discharge burning in steam-shell, which showed that increase the steam-shell starts from the tip of the electrode. In our case, in contact electrode with the surface of the glass, increase steam-shells from the electrode impossible. As a result, was observed instability discharge burning due to failure of the steam-formed shell. A similar course of plasma-electrolytic process takes place by using the pulsating voltage. It showed the need for discharge ignition with no contact with the glass and the successive approximations of electrode to the glass surface.

By using the above arguments, experiments were conducted to glass treatment and polyethylene at the anodic and cathodic polarization of the electrode.

Experiments with a metal cathode showed that there are two modes of processing the glass. These modes differ in terms of removal rate of the glass surface, the first mode is characterized by a speed of 100 microns/s was observed near the surface to a depth of 350 - 400 microns. The second mode is characterized by a decrease in the rate of removal of the glass material to 10 microns/sec is observed at a depth of 400 microns. This observation coincides with research presented in [7]. Removal mechanism proposed in this work is based on two phenomena of melting glass under the action of thermal energy of the discharge and etching glass with sodium hydroxide. It is assumed that the heat from the discharge locally increases the surface temperature of the glass, thereby causing it to melt and to evaporate the water from the treatment zone. In the processing zone remains NaOH as the salt melt. It is assumed that the OH radicals are etched glass with the following reaction:

$$2\ NaOH + SiO_2 \rightarrow NaSiO_3 + H_2O$$

$NaSiO_3$ should be washed away due to the electrolyte. But this is contradicts the previous approval of the evaporation of fluid from the treatment area. Analysis of the resulting glass surface after of plasma-electrolytic treatment showed that at the edges of the hole there are the remnants of molten glass material, which speaks in favor of the thermal melting glass and extruded products from the treatment zone melting electrode-cathode under the influence of mechanical loading. In the case of treatment discharge with the metal anode was found that, in contact vapor-gas envelope with surface of the glass discharge burning at the boundary of three phases: glass - electrolyte - vapor-shell. Gradual approximation metal anode to the glass has resulted in discharge burning between the anode and glass in vapor-shell what is not observed in the cathode regime. Discharge burning between the anode and the glass was accompanied by an intense glow. In this case, destruction of the glass surface occurs under the influence of plasma etching. The etching rate was 140 microns/s. Investigations on the treatment of polyethylene showed a completely different picture. Discharge burning occurs only between the metal anode and the electrolyte cathode. Figure 3 shows a photograph of the slice the interelectrode gap. It clearly shows that the streamers appear at the tip of the electrode and evolve on the electrolyte surface (steam-border membranes). Since the thermal capacity of of plasma-electrolytic discharge with metal anode ten times less than with a metal cathode, and consequently the processing performance of polyethylene was very low.

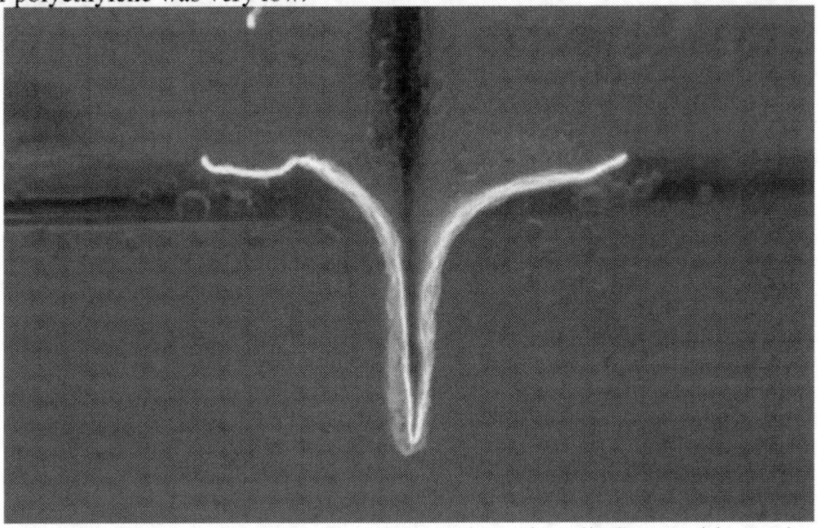

Fig. 3 Slice of the interelectrode gap at plasma-electrolyte discharge with metal anode.

The research resulted in found that the mechanisms of removal of glass for anodic and cathodic regimes differ. At cathodic regime occurs mainly thermal effect from the discharge. And at anodic regime occurs mainly glass etching by plasma of discharge.

Acknowledgments
The authors gratefully acknowledge the financial support from the Russian Foundation for Basic Research №13-08-00906 A.

References

[1] Kurafuji H and Suda K. 1968 Electrical discharge drilling of glass Ann. CIRP 16 415-9.
[2] Jalali M, Maillard P, Wutrich R. Toward a better understanding of glass gravity-feed micro-hole drilling with electrochemical discharges. J. Micromech. Microeng. 19 (2009) 045001 (7 pp)
[3] Mousa M, Allagui A, Ng H D, Wutrich R. The effect of thermal conductivity of the tool electrode in spark-assisted chemical engraving gravity-feed micro-drilling. J. Micromech. Microeng. 19 (2009) 015010 (7 pp).

[4] Jana D. Abou Ziki, Rolf Wüthrich. Forces exerted on the tool-electrode during constant-feed glass micro-drilling by spark assisted chemical engraving. International Journal of Machine Tools & Manufacture 73 (2013) 47–54.

[5] Kashapov L, Kashapov N, Kashapov R. Research of plasma-electrolytic processes in various ratio of the area anode to the cathode.

[6] Kashapov R. Research of plasma-electrolytic processes of formation microrelief surfaces of metals. Dissertation for the degree of . 2012.

[7] Jana D, Abou Ziki, T Didar, R Wuthrich. Micro-texturing channel surfaces on glass with spark assisted chemical engraving. International Journal of Machine Tools & Manufacture 57 (2012) 66–72.

Investigation of the process of plasma-electrolyte formation surface microrelief of cobalt chromium alloy

L N Kashapov[1], N F Kashapov[1] and R N Kashapov[1,2]
[1] Engineering Institute, Kazan Federal University, Kazan, 420008, Russia
[2] Laboratory of Radiation Physics, Kazan Physical-Technical Institute of the Kazan Scientific Center of the Russian Academy of Sciences, Kazan, 420029, Russia

E-mail: kashramiln@gmail.com

Abstract. The goal is to investigate the possibilities of plasma-electrolytic formation of microrelief for replacement method of sandblasting. We found that with the cathode mode of plasma electrolytic surface treatment, CoCr-alloy has two kinds of structures: "porous" and "reflow". "Reflow" the surface was also covered with tubercles, the size of 200 - 300 nm. Analysis of roughness parameters and surface microrelief showed the possibility of replacing the sandblasting on the plasma-electrolytic treatment.

1. Introduction

The basis for the development of technology and industry is to develop new economic and environmentally friendly methods of formation of the products and their subsequent processing. A special place is occupied by surface finishing methods, aimed at giving special performance properties. One of these widespread methods is sandblasting. This method has several disadvantages, namely the creation of a large amount of dust, which must be removed with special suction devices, the gradual destruction of the nozzle, the flow of sand, the presence of micro and nanoparticles of aluminum oxide on the metal surface, impair the adhesion of the deposited coating. This causes the need to find a new method of formation microrelief not having these drawbacks. Recently, increasing distribution began receiving plasma-electrolytic processing method [1,2,3]. One of its features is the ability to form a microrelief surface of metals [4,5]. The purpose of this work was to investigate the process of formation of surface microrelief cobalt chromium alloy for further replacement sandblasting. CoCr alloy was chosen because of its widespread use in the manufacture of dental prosthesis carcass. The presence of nanoparticles of aluminum oxide on its surface, resulting flaking of ceramics.

2. Experimental

In industry mainly used product obtained by the method of machining or by casting processes. In our work, for the approximation to the real conditions of the experiment samples from a cobalt chromium alloy have been cast on the fore-vacuum casting installation in the form of flat plates measuring 5 * 4 * 2 mm. For research was selected cobalt chromium alloy firm Starbond CoS with the following composition and characteristics: Co - 59%, Cr - 25%, W - 9%, Mo - 3,5%, Si - 1%, another components (C, Fe, Mn, N) - max 1,5; ultimate tensile strength (Rp 0.2) 650 MPa, tensile strength 910 MPa, ultimate elongation 8, modulus of elasticity 200 GPa, Vickers hardness 280 HV 10, density 8,8 г/см³, solidus-liquidus interval 1305 - 1400°C, casting temperature 1500 - 1550°C, thermal expansion coefficient 14,0 μm/м °C.

Beforehand was conducted mechanical treatment of the samples by carborundum disk. The surface morphology was investigated by scanning electron microscope EVO 50 Carl Zeiss AG. The elemental composition was determined by electron probe microanalysis using X-ray microanalysis system INCA Energy and INCA Wave. Profilogram surface roughness measured using a profilometer TR-200.

For plasma-electrolytic formation of microrelief was used experimental setup [4]. We investigated the influence of ripple voltage. To prevent resizing the area of the anode dielectric was wearing a protective cover, which leaves open only strictly defined portion of the electrode. Anode area was 20 cm^2, cathode area - 0.12 cm^2. Electrolytic bath filled with electrolyte - aqueous solution of sodium hydroxide concentration of 5% by weight. The main operating parameters of the electrode system are: the voltage across the electrodes, the magnitude of the discharge current, the current density at the electrodes. Changing the form of voltage and current at the time of discharge ignition was determined by an oscilloscope FLUKE 105 SCOPEMETER SERIES II, timebase varied from 5 ns to 60 s. Measuring the voltage at the anode via a digital multimeter APPA 105, the relative measurement error is 0.1%.Using a DMM APPA 305 we measured the temperature of the electrolytic cathode by using chromel-copel thermocouple. The electrolyte temperature before and after the experiment were measured with a mercury thermometer.

Figure 1 shows the surface morphology of the samples after casting and final polishing. As can be seen the surface is flat and has a value of roughness R_a = 0,5 micron.

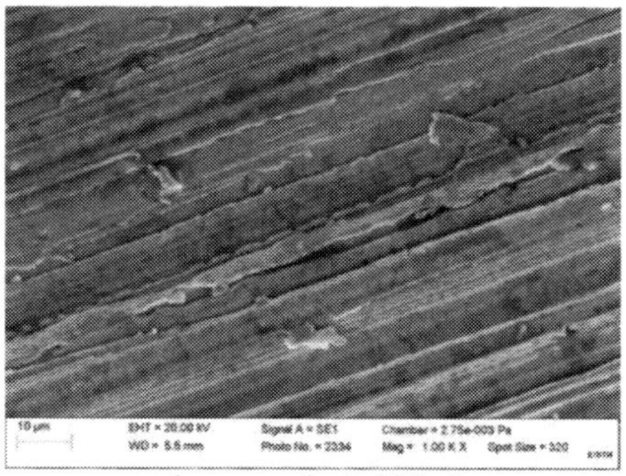

a – increase in 1,000 times

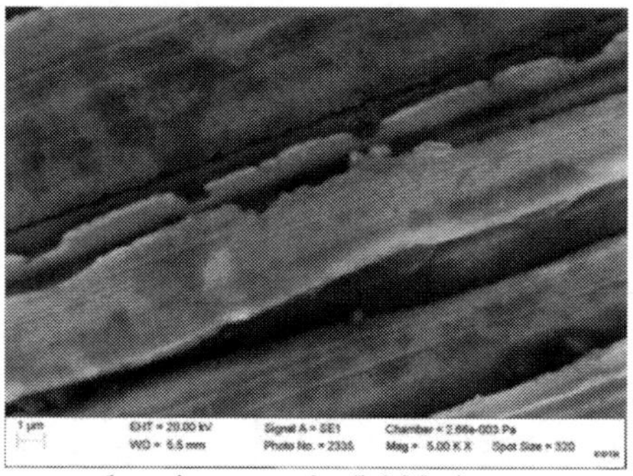

b – increase in 5,000 times

Fig. 1. SEM images of the surface after treatment carborundum disk.

Plasma electrolytic treatment was carried out in two ways, first, when the discharge was ignited at the anode, and the second, when the discharge was ignited at the cathode. Required roughness parameters are not achieved in the anodic treatment mode, a gradual dissolution of the metal. Treatment in the cathode mode enabled to form microrelief of CoCr alloy and have various structures.

Before the experiments on the formation of microrelief was obtained voltage-current characteristic of the discharge on CoCr electrode, presented in Figure 3b. The discharge was ignited at a voltage of 130 V, this point is the maximum of the curve. Next, with increasing voltage, there is a decrease of the current, but the glow discharge increases. The formation of microrelief was conducted in the voltage above 160 V. After processing were obtained two groups of surfaces, the first is represented in Figures 2a, 2b and 2c, the second - 2d, 2e and 2f. The first group we have called "reflow" surface, so observed the formation of metal droplets of size 1 - 3 microns and their fixation on the surface. This process is observed at a voltage of 180 V. The second group we have called "porous" structure, so the surface is porous. In this mode is not observed the formation of droplets of molten metal and formed pore size of 4 - 5 micrometers.

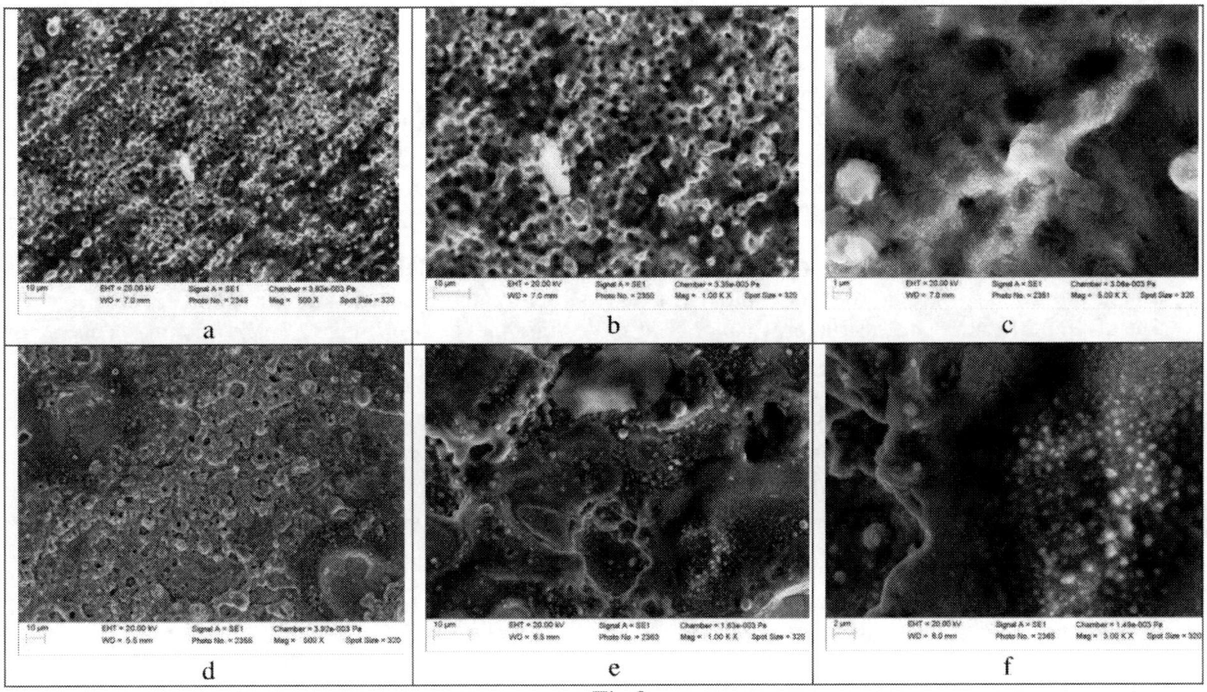

Fig.2

Figure 3a shows profilogram of surfaces. On it were defined the main parameters of roughness: R_a = 1,69, R_y = 12,25, R_t = 15,43, R_m = 6,88, R_z = 8,43. The resulting surface has all the necessary parameters for the effective retention of the microrelief applied ceramic coating.

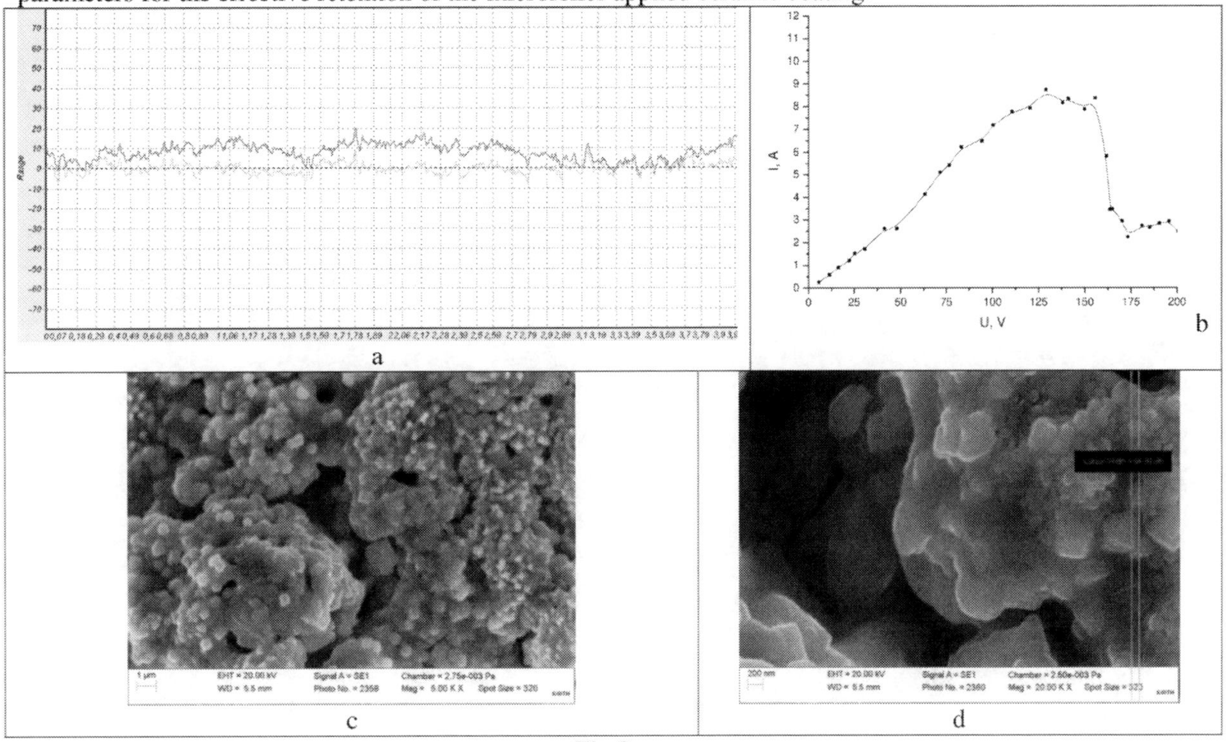

Fig.3

A more detailed analysis of the "porous" structure showed that the edge of pore consist of mounds with a typical size of 200 - 300 nm. It is shown in Figures 3c and 3d. Formation of porous structure occurs at a discharge voltage of 195 V. When burning, discharge is intense heat and prolonged exposure at high voltages can lead to melting of the electrode. Thus important parameter is the processing time.

3. Conclusions

Analysis of the plasma-electrolytic treatment shows that it is possible to obtain several types of structures on the surface of the CoCr alloy: "melted" structure and "porous" structure, covered spurs size of 200-300 nm. These structures have an average roughness parameter R_a = 1.7 microns. This coincides with the roughness parameter is obtained after a standard method of sandblasting. But considering the porosity and the presence nanoshpor on the surface of the metallic carcass after plasma-electrolytic processing is assumed that the value of the adhesion of ceramics to increase because the value of specific surface area increases.
Thus, the application of the method of plasma-electrolytic treatment would eliminate the presence of dust on the surface of the metal and increase the adhesion of the ceramic to the metal surface.

Acknowledgments

This work was funded by the subsidy allocated to Kazan Federal University for the state assignment in the sphere of scientific activities.

References

[1] A. Allagui, E. Baranova ,R. Wüthrich. Electrochimica Acta 93 (2013) 137– 142
[2] R. Wüthrich, A. Allagui, Electrochimica Acta 55 (27) (2010) 8189
[3] S. Sato, K. Mori, O. Ariyada, H. Atsushi, T. Yonezawa, Surface and Coatings Technology 206 (5) (2011) 955.
[4] L. Kashapov, N. Kashapov and R. Kashapov, Journal of Physics : Conference Series. 479(1), 012005, 2013
[5] L. Kashapov, N. Kashapov and R. Kashapov, Journal of Physics : Conference Series.479(1), 012003,2013.

Steel surface modification with plasma spraying electrothermal installation using a liquid electrode

A A Khafizov[1], R I Valiev, Yu I Shakirov, R A Valiev

Branch of Kazan Federal University in Naberezhnye Chelny, 68/19 Mira Street, Naberezhnye Chelny, 423810, Russia

E-mail: Almazok75@yandex.ru

Abstract. Recently, much attention has been paid to different processes using low-temperature plasma, and in particular, the process of plasma spraying. Despite the fact that the plasma spraying method has been established for a relatively long time, there are several unsolved issues in this field that are associated with the choice of the optimal spraying modes. It is connected with the fact that the development of optimal spraying process modes is a rather difficult task, since the problem of creating an optimal design for the plasmatron is not solved yet. In this article the technological plasma plant with liquid electrode is discussed, which provides a plasma spray with a temperature up to 5000°C and lengths up to 100 mm. Engineered installation allows to carry out plasma spraying of steel surface. The process and parameters of plasma spraying optimal modes are examined in the article.

1. Introduction

The hardening and corrosion protection problems of new-produced parts and recovery problem of worn parts surfaces are the problems of today, because the industry is forced to replace complex, labor-consuming and expensive new parts and connection joints for the further machines operation until complete recourse has being produced [1].

One of the most perspective ways of hardening and protecting the parts surfaces is plasma spraying [2, 3], especially for machine parts operating at high temperatures. Deterioration can reach 1 – 1.5 mm. As these parts are complex-alloyed, the conventional restoration methods of welding, overlaying and electroplating are prohibitive because of defects in the weld zone, the impossibility of welding, distortion, structural changes in the metal, the duration of the deposition electroplating process, , etc. [4, 5]. All these disadvantages are absent during surfaces spraying: it can be applied to various refractory materials, it can be sprayed on different materials, the coating can be applied both to large and to limited sections, the possibility of spraying the coating thickness of 2 mm and multilayer coatings, basic light heating , no structural changes in basic, equipment simplicity [6].

2. The experimental installation

Functional diagram of the experimental electrothermal installation (Fig. 1) intended to produce electric charge [7] at atmospheric pressure consists of a power supply system, equipment for storing, cleaning, feeding and regulating electrolyte; changeable electrolytic cells – basins; installation operation monitoring and control equipment and apparatus for measuring the characteristics of

[1] To whom any correspondence should be addressed.

Journal of Physics: Conference Series **567** (2014) 012026 doi:10.1088/1742-6596/567/1/012026

electrical discharges, thermal processes parameters and effects of plasma discharge electrolytic electrode on solids.

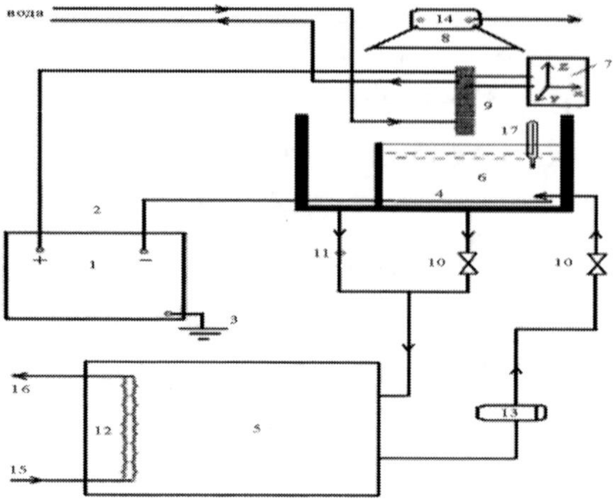

Figure 1. Functional chart of the experimental electrothermal installation
1 - power supply; 2 - current supply; 3 - grounding switch; 4 – current-carring plate; 5 - container for electrolyte storing; 6 - electrolytic cell; 7 - pointing device; 8 - exhaust hood; 9 - solid metal electrode; 10 - valves; 11 - check valve; 12 - coil; 13 - pump; 14 - fan; 15, 16 - pipe cooler; 17 - thermometer.

Electric discharge occurs between the liquid and the solid (metal) electrodes in a range of interelectrode distances $l = 1 \div 100$ mm, the current $I = 10 \div 100$ A, the voltage $U_P = 40 \div 300$ V. As electrolytic electrode process water, mortars of NaCl, $CuSO_4$ of different concentration, etc. were used.

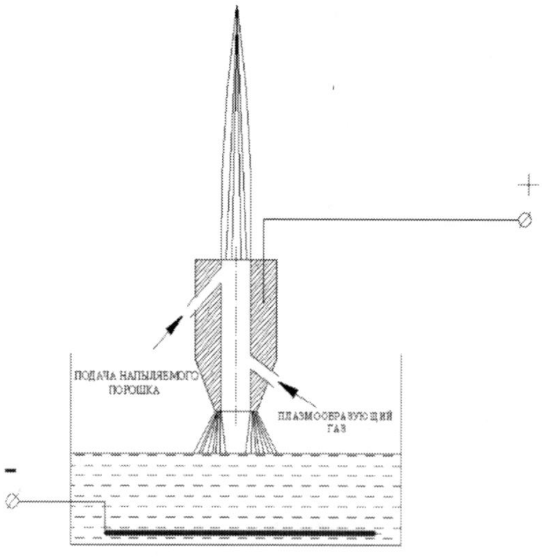

Figure 2. A device with the discharge chamber in the form of a truncated hollow cylinder for producing the plasma jet

The workflow where a powder chamber in the form of a truncated hollow cylinder is used as a metal electrode is shown in Fig. 2. The electrical discharge is produced between the bottom edge of the metal and the flow of the electrolytic electrodes. When the metal electrode surface temperature reaches a certain value (about 50°C) plasma jet occurs through the interior channel of the discharge chamber. The length of this jet essentially depends on the discharge current I, the interelectrode distance l, the type of the discharge chamber and the flowing electrolyte and metal electrode temperature. To maintain optimum surface temperature of the metal electrode cooling with tap water is used.

3. Results and their discussion

Gas (argon, nitrogen, air) passes through a burning arc between two electrodes (the electrode, the electrolyte and the copper nozzle). Due to the high energy of the burning arc gas atoms lose electrons from their outer shells. The result is an ion-electron gas or plasma. Coming out of the nozzle copper electrode at high speed electrons fill their places in orbits around atoms. As a result, energy is released. The temperature of the plasma spray reaches 3000 – 5000°C. In the zone at the outlet of the nozzle sprayed material is supplied into the plasma spray in powder form. The iron oxide powder Fe_3O_4 was the material for the deposition. As a result the spraying material is heated to the melting accelerates and applied to the surface of the workpiece. Adhesion strength of coatings with the basic amounts 20 MPa.

In the sputtering process it is necessary to control the thickness of the sprayed layer, which for plasma coating is 0.1 – 2 mm. Parts surfaces which are not to be plazma sprayed, must be protected by the special protective equipment.

Plasma coatings studies on wear resistance were conducted at a temperature of 800°C, and showed that plasma coatings have wear resistance of 1.5 – 3 times higher than the wear resistance of the substrate made of special steels and alloys.

The effect of coating thickness on the coating strength of the iron oxide in the separation of the underlayer is shown in Fig.3. Reduced adhesion of coatings with increasing thickness is due to the accumulation of residual stresses in sequential deposition of layers.

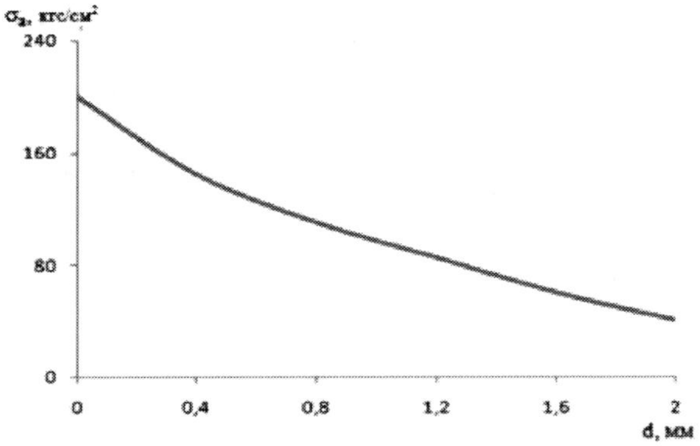

Figure 3. The adhesion strength dependence on the iron oxide coating thickness

When spraying the powder Fe_3O_4 on the thickness of 1.0 – 1.5 mm, the influence factors of the coating process on the cleavage strength are recognized. With plasmatron current increasing, the strength decreases due to increased oxidizability of the coating material and the base and residual stresses sharp increase. The increasing of the distance and the iron oxide particles flight time increases the strength. The spraying distance reduction less than 50 mm is unpractical because of overheating of the sprayed coating and peeling. The plasmatron power increasing improves bond strength due to the increase in enthalpy of the particles.

An important factor determining the strength of adhesion of plasma coatings and other physico-mechanical properties is the quality of atomization of the sprayed particles. With powder materials plasma spraying, the intensive heat transfer of the powder particles, which are in the plasma flow is occured, but the residence time of the particles in it are limited. Since the propagation of heat in the particles of the material deposited by the plasma is going on due to the thermal conductivity, the particles dimensions of the sprayed powder are important for adhesion. Large particles do not have time to melt during the residence time in the plasma spray and impinge on the surface at weakly plastic state, thereby worsening the quality of the coating. Fine particles impede powder supply into the nozzle, clog the nozzle and are rapidly cooled on the way to the product losing speed.

The optimum size selection of the powder depends on many conditions. The particle diameter can be obtained from the heat equation [8] $\partial T/\partial \tau = a d^2 T / dx^2$, wherein $a = \lambda / c \cdot \gamma$ – the thermal diffusivity coefficient; λ – thermal conductivity; c - specific heat; γ - density; τ – heating time of the powder.

The solution of the equation in the criterion form

$$(T_m - T_1)/(T_m - T_2) = f(F_0),$$

where T_m – the particles melting temperature; T_1 – the particle center initial temperature; T_2 – the final temperature of the particle center; F_0 - Fourier criterion $(F_0 = 4a\tau / r^2)$; r - the radius of the particle.

The decision criterion equation at $T_2 = 0{,}9T_m$; $f(F_0) = 0{,}1$; $F_0 = 0{,}3$ gives the maximum value of the powder diameter

$$d = 4\sqrt{a\tau / 0{,}3}. \qquad (1)$$

The calculation of the equation (1) for iron oxide gave the diameter value of 100 μm.

Granulated powder composition is to be uniform. With a wide range of particles dimensions the quality and adhesion of the coating deteriorate significantly. This is due to the uneven heating of different granulation powder.

Spraying angle affects the adhesion strength. At angles close to 90°, the adhesion strength is deteriorating due to the reduction of particle velocity by the reflected airflow. At angles smaller than 40-60°, there is a reflection of the particles of the sputtered material.

The results of measuring the peel strength of plasma coatings deposited (sprayed) at different angles are shown in Fig.4.

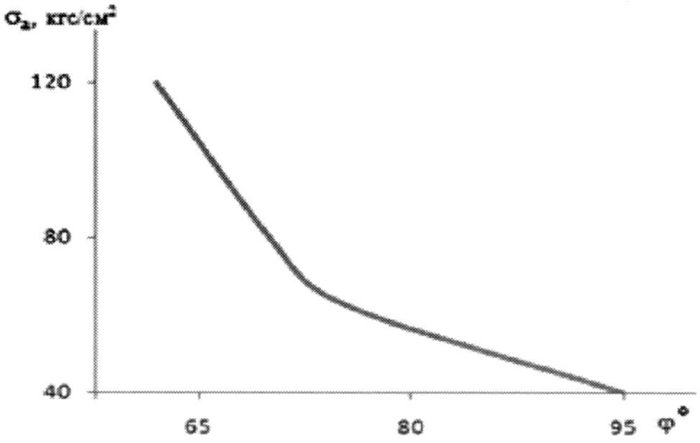

Figure 4. The adhesion strength dependence on the spraying angle

The greatest strength of adhesion of iron oxide coatings is achieved by applying an angle 60°. This is due to the fact, that with the angle decreasing from 90° to 60°, the resistance of the gas spray decreases, and consequently increases the speed of the particles by collision with the surface, which leads to a better adhesion of coatings to it. As seen from the figure, the adhesion coating strength at a spray angle of 60° 2.5 times higher than at 90°. In plasma spraying angle of less than 45° is unacceptable, because of the particles large slip, and hence, poor fixing them on the surface.

To optimize the process of plasma deposition of iron oxide on the metal substrate, the regression dependence of the adhesion strength on certain input parameters is determined. To do this, the method of planning multifactor experiment was used [9]. Since it is impossible to take into account all of the input parameters in the study, the most important parameters: the power of the discharge current (I), the magnitude of distance from the powder inlet to the substrate (h), the plasma gas flow rate (Q), and processing time (τ).were chosen for a selected sputtering method and installation. The following parameters: d - coating thickness and ρ - density of the coating were selected as the output value of optimization.

Mathematical treatment of the experimental results by using the least squares method [10] yielded the regression equation in the following form:

$d = y = 50{,}69 + 3{,}31x_1 - 5{,}22x_2 + 6{,}16x_3 + 5{,}78x_4 - 4{,}34x_1x_4 - 0{,}18x_2x_3 + 7{,}31x_3x_4$

$\rho = y = 2{,}92 + 0{,}28x_1 + 0{,}27x_3 - 0{,}34x_4 + 0{,}29x_1x_4 - 0{,}14x_2x_4 + 0{,}11x_3x_4$

Thus, the function (dependences) obtained can be used to determine the desired operating mode, as well as for process control.

4. Conclusions

These results suggest that to improve the adhesive strength comprehensive theoretical and experimental studies that allow to reveal the mechanism of a durable adhesive contact formation are needed. The use of multivariate experiments planning methods enables to optimize the deposition process in order to obtain maximum adhesion strength.

References

[1] Suslov A.G. The quality of machine parts surface layer.- M .: Mechanical Engineering, 2000. - 320 p.

[2] Rykalin N.N., V.V. Kudinov Low-temperature plasma and its application. Seminar in Moscow. Vestn. USSR Academy of Sciences, № 7,1973.

[3] Puzryakov A.F. Theoretical basis of plasma spraying technology. tutorial, 2003. - 360 p.

[4] Chernoivanov V.I. Restoration of machine parts.M.: GOSNITI, 1995. - 280 p.

[5] Peshkov V.V., Petrenko, V.R. Chemical-thermal, electrophysical treatment of metals, alloys and electroplating coatings [electronic resource]: Monograph / V.N. Gadalov, V.R. Petrenko, V.V. Peshkov, S.V. Safonov. - M: ARGAMAK - MEDIA, 2013. - 320 p.

[6] Lashchenko G.I. Plasma hardening and spraynig.- K .: "Ekotehnologiya", 2003.- 64 p.

[7] Y.I. Shakirov, R.I. Valiev, A.A. Hafizov, G.Y. Shakirova. Multichannel plasma system with electrolytic cathode. Scientific - technical magazine "Automotive industry". Moscow, 2011, № 2, p.36-38.

[8] Tikhonov A.N., Samarcky A.A., Mathematical Physics equations, the 3rd edition., Moscow, 1966.

[9] Kostenko E.M. Enumeration of multifactorial experiment standard plans/ E.M. Kostenko // News of Poltava State Agrarian Academy. - Poltava: PDAA.- 2013. -№1. – p.146-150.

[10] Adler Y.P., Markova E.V., Gramovsky Y.V. Experiment planning by searching the optimal conditions. - Moscow: Nauka, 1976. - 279 p.

Reactive magnetron sputtering model at making Ti-TiO_x coatings

A G Luchkin and N F Kashapov
Kazan (Volga Region) Federal University, 18 Kremljovskaja str., Kazan 420008, Russian Federation

E-mail: AGLuchkin@kpfu.ru

Abstract. Mathematical model of reactive magnetron sputtering for plant VU 700-D is described. Approximating curves for experimental current-voltage characteristic for two gas input schemas are shown. Choice of gas input schema influences on model parameters (mainly on pumping speed). Reactive magnetron sputtering model allows develop technology of Ti – TiO_x coatings deposition without changing atmosphere and pressure in vacuum chamber.

1. Introduction

Low-temperature plasma is widely used for surface modification and making functional coatings [1-4]. Magnetron sputtering of metal target at work and reactive gases mixture atmosphere has its difficulties such as unstable current-voltage characteristic [5]. There are several models of reactive sputtering process [6-9]. But they are not universal and for concrete experimental plant are needed in specific approximating equations system.

The aim of the paper is to create mathematical model of reactive magnetron sputtering for experimental plant VU-700D.

2. Experimental plant and model

Functional schema of experimental plant, discharge parameters are discussed in [10,11]. Metal Titanium target 100x500 mm^2 was used. Experimental plant provides gas input in region of magnetron target (schema “A”) and in region of substrate (schema “B”). Volt-ampere characteristics measured within input of work and reactive gases in vacuum chamber at different ratio. Voltage fixed at different values of discharge current.

Reactive magnetron sputtering model has several assumption: reactive gas (oxygen) is ideal diatomic gas; at the same time oxygen is pumped out, adsorbs on vacuum chamber walls and magnetron target and desorbs from magnetron target; there is oxide film on magnetron target surface and ratio of magnetron target surface covered by oxide film to total magnetron target surface is θ ($0< \theta <1$); $\theta = \theta(r)$, where r – magnetron target radius. At stationary state for target we can write [9]:

$$\sigma kp(1-\theta(r)) = \frac{\eta_0 j(r)\theta(r)}{e} + \frac{N\theta(r)}{\tau(T)}, \quad (1)$$

where $\sigma = 1$ – oxygen adsorption coefficient, $k = 3{,}81 \cdot 10^{18}$ $sm^{-2}sec^{-1}Pa^{-1}$ – quantity of oxygen atoms hits on the surface at 300 K, η_0 – sputtering coefficient, e – electron charge, $j(r) = j_0 f(r)$ – current density. $N\theta(r)/\tau(T)$ – oxygen thermal desorption, N – oxygen desorption centers density, $\tau(T)$ – oxygen presence time on target surface with temperature T. We assume that:

$$\frac{\eta_0 j(r)\theta(r)}{e} \gg \frac{N\theta(r)}{\tau(T)}. \quad (2)$$

Material sputtering speed:

$$R = \eta_m \frac{I}{e} = \frac{\eta_m}{e}\int_{r_1}^{r_a} j(r)\big(1-\theta(r)\big)2\pi r\, dr, \quad (3)$$

where η_m – material sputtering coefficient. Sputtered material homogeneously deposits on getter surface F, where oxidation degree is ϑ $(0 < \vartheta < 1)$. Oxygen partial pressure in chamber depends on gas input volume D_0, pumping speed S_p and sputtering speed R:

$$p = \frac{D_0}{S_p} - \frac{\gamma\vartheta R}{S_p}, \quad (4)$$

where γ – factor is about $k_B T$. For ϑ can be written [12]

$$\vartheta = \frac{1}{2}\frac{kp(1-\vartheta)}{R/F}, \quad (5)$$

where adsorption coefficient $\sigma = 1$ and there is no desorption.

From (5) follows:

$$\vartheta = \frac{\frac{1}{2}kpR/F}{1+\frac{1}{2}kpR/F}. \quad (6)$$

The insertion of (6) into (4) gives

$$S_p p = D_0 - \gamma R \frac{\frac{1}{2}kpR/F}{1+\frac{1}{2}kpR/F}. \quad (7)$$

From equation (1) define $k\,p$ and insert it into (7). After conversion we will have

$$\frac{S_p}{k}\left[2R + \frac{\theta(r)}{1-\theta(r)}\frac{\eta_0}{e}j(r)F\right] + \gamma FR = D_0\frac{1-\theta(r)}{\theta(r)}\frac{e}{j(r)\eta_0}\left[2R + \frac{\theta(r)}{1-\theta(r)}\frac{\eta_0}{e}j(r)F\right]. \quad (8)$$

From equations (3) and (8) discharge current I could be defined as function of $\theta_0 = \theta_0(r)$ $(0 < \theta_0 < 1)$

$$I(\theta_0) = \frac{B_2 D_0 \cdot e \cdot \left(\frac{1}{1-\theta_0} + B_1 \frac{2\cdot A_t \cdot \eta_m}{\eta_0 \cdot F}\frac{(1-\theta_0)}{\theta_0}\right)}{\frac{S_p \cdot \eta_0}{k \cdot A_t}\frac{\theta_0}{(1-\theta_0)^2} + B_1 \cdot \eta_m \cdot \left(\gamma + \frac{2\cdot S_p}{k \cdot F}\right)}, \quad (9)$$

where integrals

$$B_1(\theta_0) = \frac{1}{A}\int_{r_1}^{r_a}\frac{f^2(r)2\pi r\, dr}{\theta_0 + (1-\theta_0)f(r)} \quad (10)$$

and

$$B_2 = \frac{1}{A}\int_{r_1}^{r_a} f(r)2\pi r\, dr \quad (11)$$

could be found from current density distribution function $f(r)$ and effective target surface A. D_0 –total gas input (argon and oxygen) [mW] or [Pa l/sec], F – getter surface square, sm^2, S_p = 6 l/sec – pumping speed, η_m and η_0 – metal and oxide sputtering coefficients.

Target oxidation degree influence on current-voltage characteristic could be written as

$$I = AC\left(\frac{U}{U'}\right)^{b+cB_3\theta}, \quad (12)$$

where integral

$$B_3 = \frac{1}{A}\int_{r_1}^{r_a}\frac{2\pi r\, dr}{\theta_0 + (1-\theta_0)f(r)} \quad (13)$$

could be found from view of $f(r)$. From equation (12) voltage expresses as

$$U(\theta_0, I) = U'\left(\frac{I}{C\cdot A}\right)^{\frac{1}{b+cB_3\theta_0}}. \quad (14)$$

Parametric representation of current-voltage characteristic is given by equations (9) and (14). For results illustration we assume $f(r) = B_1 = B_2 = B_3 = 1$. Simplified calculations qualitatively match with current-voltage characteristic curves main behavior. Other distribution functions $f(r)$ don't change results seriously.

3. Results and discussions

Experimental current-voltage characteristics for different schemas of reactive gas input and their approximation are shown at pic. 1 and 2. Solid line – approximating curve, circles and squares – experimental data, increasing and decreasing current density respectively. Lower increasing part of approximating curve corresponds to "oxide" regime, when oxide deposits on substrate. Higher increasing part corresponds to "metal" regime, when substrate surface is coated by metal film. Middle part corresponds to forbidden states.

When discharge is burning the transparent coating is making. Increasing of discharge power at defined moment oxide film at target surface sputters totally and metallic coating is making. Discharge power decreasing results to opposite effect but at lower value. Using these transitions we can deposit on substrate surface Ti – TiO_x coatings without changing atmosphere and pressure in vacuum chamber.

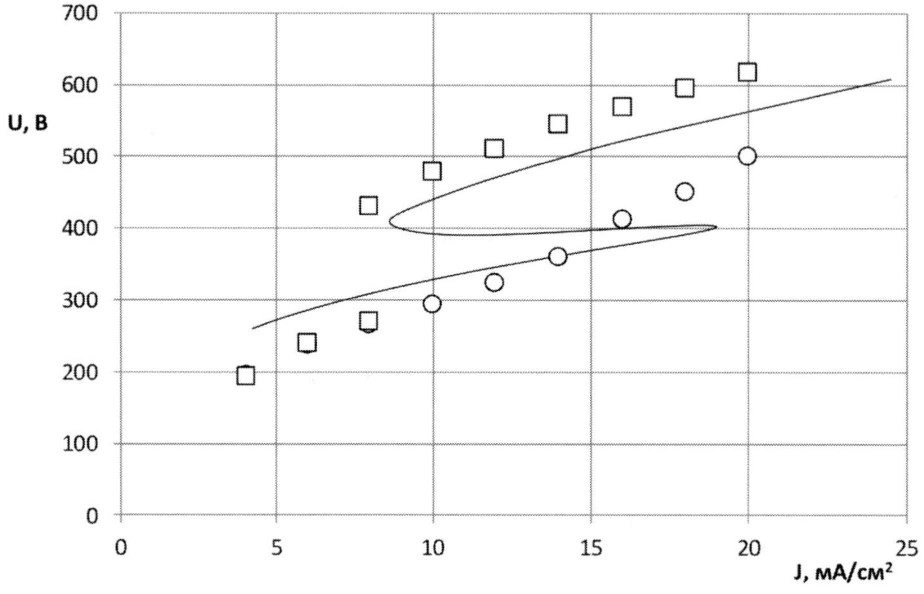

Pic. 1. Current-voltage characteristic approximation, schema "A", $4{\bullet}10^{-2}$ Pa, oxygen – argon ratio is 2 to 1: — - U(I); ○ – increasing current density; □ - decreasing current density.

Approximations for schemas "A" and "B" differs only by S_p and c. Pumping speed S_p for schema «A» grater pumping speed for schema "B" 4,5 times, coefficient c – 2 times. That difference could be explained if we remember that in schema "A" oxygen inputs directly into magnetron target region and that for oxide deposition on substrate surface responds target surface oxidation. Deviations of approximating curves from experimental current-voltage characteristics could be caused by model assumptions. For example, partial oxidation of titanium, other getter surfaces, etc.

4. Conclusions

Mathematical model of reactive magnetron sputtering for plant VU 700-D is described. Approximating curves for experimental current-voltage characteristic for two gas input schemas are shown. Choice of gas input schema influences on model parameters – pumping speed and coefficient c. Reactive magnetron sputtering model allows develop technology of Ti – TiO_x coatings deposition without changing atmosphere and pressure in vacuum chamber.

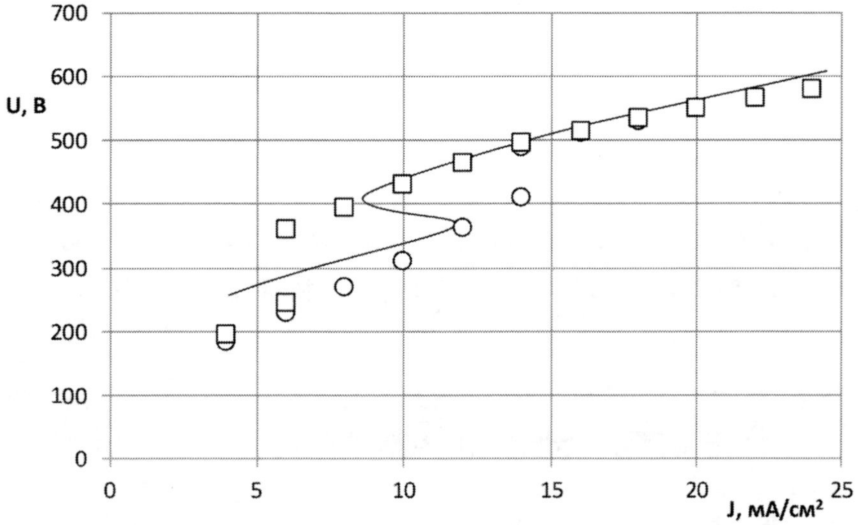

Pic. 2. Current-voltage characteristic approximation, schema "B", 4•10^{-2} Pa,
oxygen – argon ratio is 2 to 1: — - U(I); ○ – increasing current density; □ - decreasing current density.

Acknowledgements
The work is performed according to the Russian Government Program of Competitive Growth of Kazan Federal University.

References

[1] Galyautdinov R.T., Kashapov N.F., Luchkin G. S. // Welding International. - 2003.- V. 17,- № 8.- P. 655 – 658.
[2] Galyautdinov R.T., Kashapov N.F., Luchkin G. S. // Applied physics. . - 2005.- № 6.- P. 88-93.
[3] Galyautdinov A.R., Galyautdinov R.T., Kashapov N.F., // Bulletin of KGTU. Tupolev. 2010.- № 4. - P. 122-128.
[4] Galyautdinov R.T., Kashapov N.F., Luchkin G. S. // Engineering-physical journal. -2002. - V.75.- № 5. - P 170-173.
[5] Kashapov N.F., Luchkin A.G. // Russian Physics Journal 57 (2014) № 3/3, P. 162-166.
[6] S. Berg, T. Nyberg // Thin Solid Films 476 (2005) 215-230
[7] Anatoly A. Barybin and Victor I. Shapovalov Journal of Applied Physics **101**, 054905 (2007); doi: 10.1063/1.2435795.
[8] E. V. Berlin, S.A. Dvinin, L.A. Seidman Vacuum technology and technic for thin film deposition and etching M: Technosphera, 2007. — 176 pp, ISBN: 978-5-94836-134-5.
[9] Steenbeck, K., Steinbeiss, E. & Ufert, K.-D. // Thin Solid Films. 1982. V. 92. P. 371—380
[10] N.F. Kashapov, A.G. Luchkin & G.S. Luchkin. *J. Phys.: Conf. Series* **479** (2013) 012019.
[11] Galyautdinov R.T., Kashapov N.F. & Luchkin A.G. // Bulletin of KGTU. Tupolev. 2013.- № 3. - P. 106-110.
[12] Ratter, E., Monatsch Chem, 95 (1964) 795.

Journal of Physics: Conference Series **567** (2014) 012028 doi:10.1088/1742-6596/567/1/012028

Effect of supplementation of water vapor to the environmental characteristics of the combustion of propane-air mixture

V M Larionov, G A Mitrofanov and O V Iovleva,

Kazan (Volga region) federal university, Kazan, Tatarstan, Russia

E-mail: techphyskfu@gmail.com

Abstract. To improve the efficiency of combustion of fuel gas and air can be used additive steam. The article presents the results of an experimental study of the influence of water vapor on the combustion of propane-butane mixture with air. Combustion mixture produced in a modified Bunsen burner. Studies carried change of steam temperature of 180 to 260 degrees Celsius, and the change of the specific volume steam in the composition of the fuel mixture. Influence steam on combustion was estimated by the change of temperature of heating the quartz tube. It has been established that the increase of the steam temperature and increasing the specific volume of the heated vapor in the composition of the gaseous fuel increases the temperature of combustion.

Combustion efficiency can be improved in various ways.
These methods are known as methods of combustion control [1].

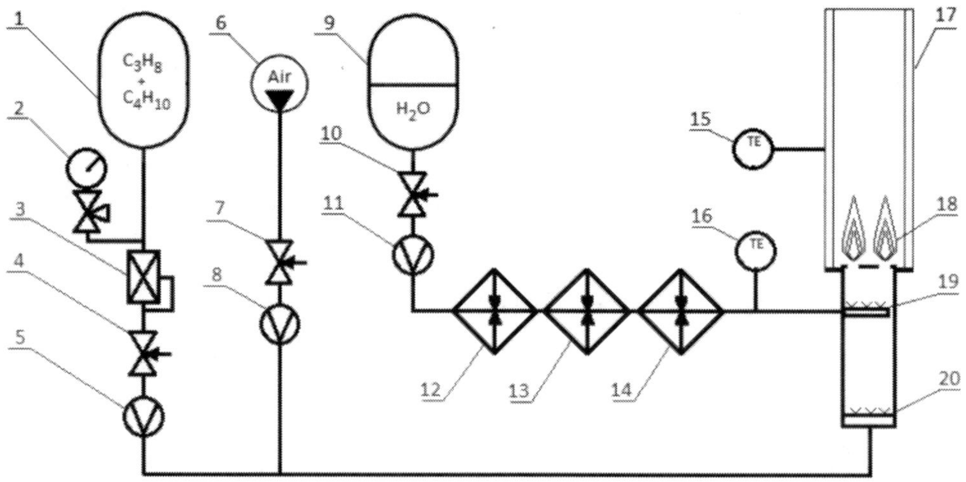

Figure 1. The main elements of the experimental stand.

Great interest in the practice of creating burners are methods based on the use of various types of fuel additives. In recent years, researchers are paying attention to water-steam additives. Experiments indicate that the presence of water vapor in the gaseous fuel increases energy efficient to install and also affects the stability of its operation [2].

The purpose of this paper is to determine the optimal characteristics of water vapor when it is used as a fuel additive when burned propane-air mixture.

The main parts of experimental plant (Figure 1) are: 1 - the fuel tank with propane-butane gas; 2 - air pump; 9 - a water tank; 12, 13, 14 - and the steam heater; 20 - Bunsen burner; 17 - silica glass. The remaining elements are the control equipment and the flow measurement components and fuel vapor, as well as temperature sensors - 15 and 16.

The temperature measurement was made on the wall of the quartz tube at a certain point with chromel-alumel thermocouple (Figure 2,a). In addition, the temperature field was estimated using the thermal imager{ Figure 2,b). Supply of heated water vapor significantly affects the nature of the flame during combustion (Figure 2,c and 2,d).

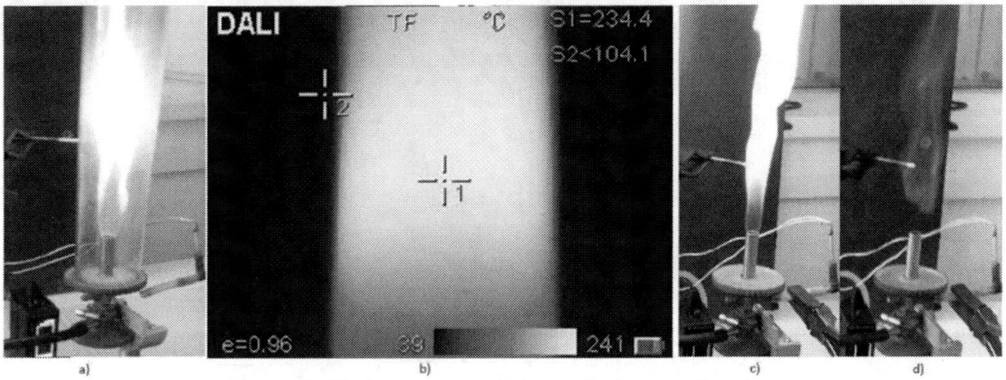

Figure 2. Various types of flame during combustion.

During the experiments measured the wall temperature of the quartz tube. This changes the temperature of the steam that enters the combustion zone. In addition, we measured the temperature of the wall of the quartz tube when changing the volume flow rate of steam at a constant temperature.

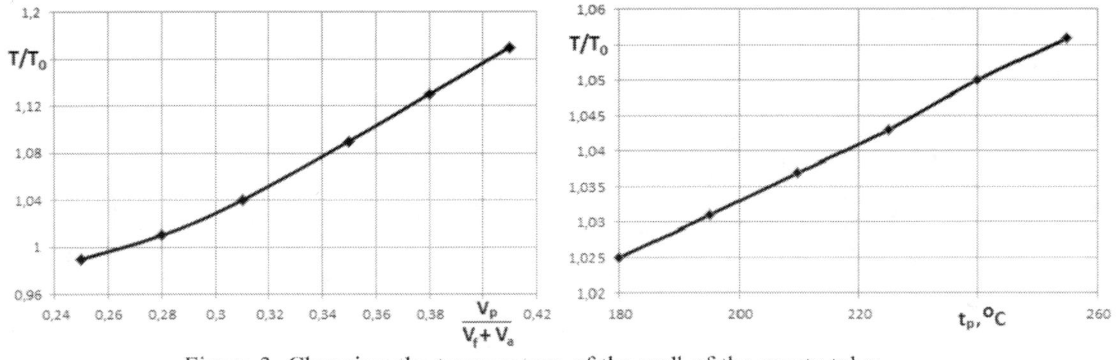

Figure 3. Changing the temperature of the wall of the quartz tube:
a) - depending on the volume fraction of steam;
b) - depending on the temperature of the steam.

Journal of Physics: Conference Series **567** (2014) 012028 doi:10.1088/1742-6596/567/1/012028

The main results of the work are presented in Figures 3a and 3b. In Figure 3a shows the experimental dependence of the relative change in temperature of the wall of the quartz tube, depending on changes in the specific volume of steam. Here, T - temperature of the wall during combustion with steam T0 - wall temperature of the quartz tube while burning without steam.

In Figure 3b shows the experimental dependence of the relative change in temperature of the wall of the quartz tube at the point of control of the temperature of heating steam.

Increasing the temperature of the steam, as well as increasing the specific volume of the steam supplied into the combustion zone, resulting in an increase in the wall temperature of the quartz tube. Consequently, the supply of the superheated steam as an additive increases the temperature of the fuel in the combustion zone.

During the experiments, we measured the concentration of CO in the flue gas. It has been found that feeding of steam concentration of CO in the combustion products is reduced by about two times.

References

[1] Ahmadullin A N, Ahmethanov E N, Iovleva O V, Mitrofanov G A 2013 *Journal of Physics, Conference Series* **479** 012004.

[2] Bykovets A P, Larionov V M, Marchukov E Yu 1992 *Izvestiya Vysshikh Uchebnykh Zavedenij. Aviatsionnaya Tekhnika* **3** 71-74

Module for dielectric surfaces modification by fast neutral particles beams.

V T Barchenko[1, 2], A A Lisenkov[1, 2], N A Babinov[1]

[1] Saint Petersburg Electrotechnical University "LETI", Saint Petersburg, Russia
[2] Institute of Problems of Mechanical Engineering, Saint Petersburg, Russia

E-mail: VTBarchenko@yandex.ru

Abstract. In this paper, we describe the module for dielectric and wide-gap semiconductor surfaces modification by fast neutral beam. The module can be used for cleaning, etching or assisting of films deposition. The surface proceeding by neutral beam can prevent an accumulation of surface charge without using current compensation by inserting electrons to the beam or RF power supply. The module beside cathode and anode contains an electrode with floating potential. Insertion of the additional electrode causes electron retention in an electrostatic trap resulting the reducing of the module operating pressure. Moreover, the electrode with floating potential allows increasing the current efficient of the module. An important feature of the module is that neutralization of the ions extracted from the plasma occurs in the cathode potential well. Thereby ions that have not neutralized cannot leave near-cathode region and there are no fast ions in the output beam. Module does not contain sources of the magnetic fields or elements heated by external sources. Module operates with free cooling. Thus, the module does not need water cooling and can be freely moved in the vacuum chamber.

1. Introduction

The processing of dielectric and semiconductor surfaces has special difficulties related to charging of the surfaces if the ion beams are used. The surface charge causes defect formation as well as reduces efficiency of the further processing. To eliminate the surface charge the following methods are used: current compensation of the ion beam, combination of ion and electron beams or using RF-plasma.

The other method of surface modification is using fast neutral particles beam. In this case there is no need to neutralize the surface charge and moreover charge-induced surface defects do not formed. Charging of the surface occurs only due to secondary electron emission. However in this case surface potential does not become too large because of returning emitted electron to the surface. Thus the maximal surface potential is determined by maximal energy of secondary emitted electrons and usually less than 100 volts.

Processing of dielectric and semiconductor surfaces using fast neutral particles beam significantly reduces processing damage compared with using ion beams or direct inductively-coupled plasma processing [1–4]. It is caused by lack of charge-induced or UV-induced defects formation.

In this regards, development of module for dielectric and semiconductor surfaces modification by fast neutral particles beam is of practical interest.

2. Module description

The module scheme is shown in figure 1. The module consists of three electrodes: cathode, anode and additional electrode. The cathode of this system is a set of parallel plates fixed at a certain distance from each other and additional electrode is hollow. An output aperture of additional electrode matches laminated cathode, the two electrodes are arranged so that they form a quasiclosed volume.

When the module installed in the vacuum chamber, the anode is grounded, negative voltage from DC power supply submitted to the cathode. Additional electrode is isolated and has a floating potential.

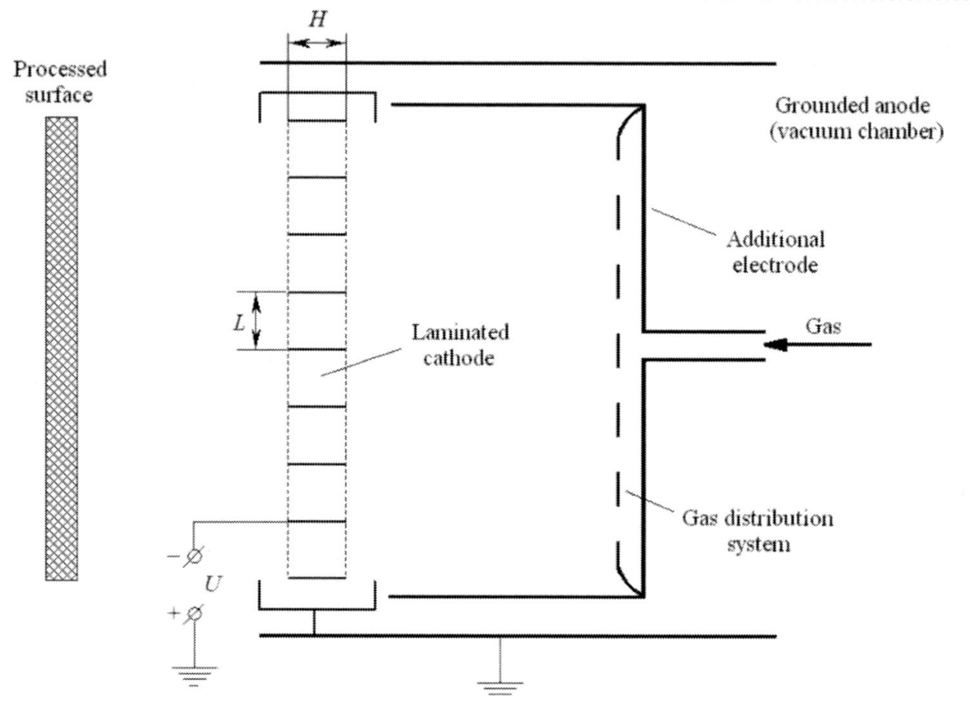

Figure 1. Module scheme

When power supply is submitted to the module, potential at the additional electrode is determined by the condition of equal electron and ion currents to the electrode. Since in the low pressure gas discharge electron temperature is significantly higher than ion temperature, there will negative potential at the additional electrode which limits electron current. As was measured the value of the additional electrode potential relative to the anode approximately is a half of discharge voltage.

The quasiclosed volume formed by cathode and additional electrode creates an electrostatic trap for fast electrons. Due to multiply reflections of electrons from the electrodes a length passed by electrons in the plasma volume before leaving it is greatly increases. At low pressure it causes reducing of the proportion of an energy carried away from the discharge by fast electrons and thus reducing of the discharge voltage. Using of the electrostatic trap allows reducing the minimal operating pressure of the module to a value of 1–2 Pa. Moreover the localization of the gas discharge plasma in the quasiclosed volume between electrodes is occurs. It allows increasing of the neutral beam flux.

Ions generated in the plasma volume are accelerated in the cathode voltage drop region. Its neutralization performed by two processes: resonant recharge on gas target and low angle reflection from cathode plates with neutralization. The percentage of the ions neutralized by each of the two processes is determined by a gas pressure in the near-cathode region and configuration of cathode plates.

Since the module is designed for operating at low pressure, the neutralization of ions by low angle reflection from metallic surface is the most important. Due to the independence of this process from the gas pressure in the neutralization region the operating pressure of the module is determined only by pressure of gas discharge maintaining.

Since the ions neutralization is occurs in the cathode voltage drop region, not-neutralized ions cannot leave the cathode potential well and thus cannot reach the processed surface with high energy. The highest possible energy of ions is determined by anode voltage drop and less than 10–20 eV. In the most cases this value of ions energy is too small to induce a defects forming. Thus, the module design provides absence of ions with high energy in the generated beam without using special devices such as separator extracting not-neutralized particles from the beam.

An electrical isolation of the additional electrode provides the zero current at this electrode. Thus, the entire discharge current flows through the cathode which plays the role of an emission grid as well. Due to this the current efficiency (the ratio of equivalent current of the beam to the discharge current) of the module is close to the 100 % and determined only by imperfect passage of the particles through the cathode. Moreover, the ion current density to the cathode and additional electrode is equal while the area of the additional electrode is significantly higher than the area of the cathode and the total ion current to the additional electrode is higher than to the cathode. Therefore there is significant secondary electron current from the additional electrode. These electrons due to the high voltage on the additional electrode have energy high enough for gas ionization. The secondary electron emission from the additional electrode noticeably reduces the discharge voltage and minimal gas pressure at which the maintaining of discharge is possible.

For additional reducing of the minimal operating pressure of the module a pressure difference between internal volume of the module and vacuum chamber is created. For this the working gas is inserted directly in the quasiclosed volume of the module through an inlet at the bottom side of the additional electrode. There is a gas distribution system between the inlet and gas discharge volume, it is required for the uniform distribution of the working gas in the internal volume of the module. Gas distributor is a plate with a plurality of holes in it. The creation of gas pressure difference under condition of constant gas flow through the module is possible due to the significant resistance of the cathode to the gas flow.

The figures 2 and 3 show photos of the module. The cathode plates is made of molybdenum to decrease energy loss on ions reflection with neutralization. All other metallic parts of the module are made of nonmagnetic stainless steel.

Figure 2. Photo of the module

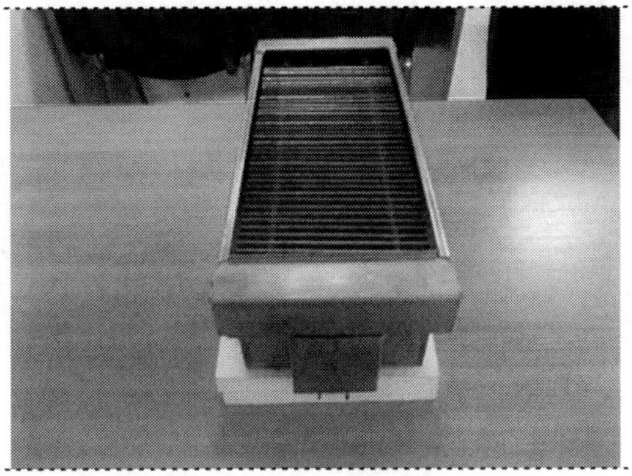

Figure 3. Photo of the module

Since there are no magnetic field sources in the module it is capable to operate while in vacuum chamber with free radiation cooling. Thus there is no need in forced water cooling and the module can be moved freely in the vacuum chamber.

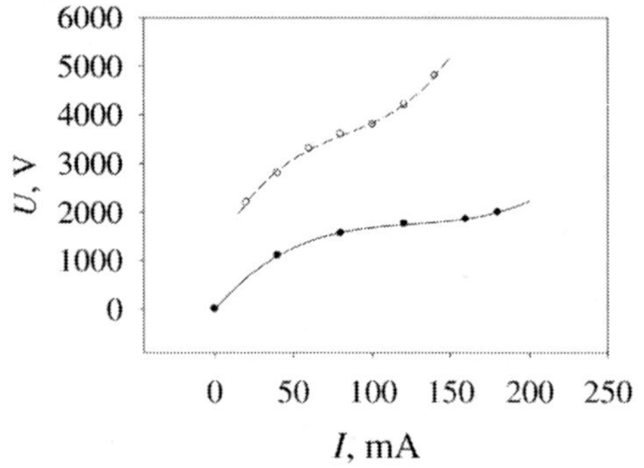

Figure 4. *I-V* characteristics of the module at different gas pressure in the vacuum chamber

The figure 3 shows *I-V* characteristics of the module at two pressures in the vacuum chamber: 3.5 Pa (black circles) and 2 Pa (white circles). The characteristics show that the module is able to operate at the gas pressure of 2 Pa and discharge current of 150–200 mA.

References

[1] Low angle forward reflected neutral beam source and its applications / B. J. Park, S. W. Kim, S. K. Kang and oth. // J. Phys. D: Appl. Phys. 2008. 41 024005

[2] Damage-free surface treatment of carbon nanotubes and self-assembled monolayer devices using a neutral beam process for fusing top–down and bottom–up processes / Samukawa S., Ishikawa Y., Okumura K and oth. // J. Phys. D: Appl. Phys. 2008. 41 024006

[3] Removal of Aspect-Ratio-Dependent Etching by Low-Angle Forward Reflected Neutral-Beam Etching / D. H. Lee, B. J. Park, G. Y. Yeom and oth. // J. of the Korean Ph. Soc. 2005. Vol. 46, № 4, pp. 867–871

[4] Effect of neutral beam etching of p-GaN on the GaN device characteristics / B. J. Park, K. S. Min, H. C. Lee and oth. // J. Vac. Sci. Technol. B. 2007. Vol. 25, № 2. pp. 295–298

Study on parts ion-plasma nitriding

I H Israphilov, V V Zvezdin, D I Israphilov, M A Chernova
Department of Energy and Information, Naberezhnochelninsky Institute (branch)
Kazan (Volga Region) Federal University, pr. Mira 68/19, Naberezhnye Chelny, 423810, Russian Federation

E-mail: DIIsrafilov@kpfu.ru

Abstract. Nitriding is used in various industries to improve the operational reliability of a wide range of products: cylinder liners, crankshafts and camshafts, spindles of machine tools, stamping and cutting tools, turbine parts, etc. Durability and performance of instruments is largely determined by the surface condition and its properties. One of the most effective ways of hardening the surface is ion-plasma nitriding.

1. Introduction

The physical nature of the processes of ion-plasma nitriding atmospheres same. However, qualitative and quantitative characteristics of the individual stages of these processes have their differences that affect the structure and kinetics of formation of the nitrided layer.

Ion nitriding process in hydrogen-containing media (ammonia, a mixture of nitrogen and hydrogen) is well studied in domestic and foreign literature [1 - 5]. Hydrogen, being a good reducing agent, has a significant effect on the kinetics of formation of the nitrided layer, intensifying the process. However, as noted previously, the presence of hydrogen in a saturating medium ko¬lichestvah significant causes embrittlement of the surface (particularly the sharp edges) and softening base under certain conditions, leads to a reduction in strength and durability of the structural elements. Addition of hydrogen in argon saturated environment enhances the ductility of the nitrided layer [2, 5]. Substitution of hydrogen with argon saturating medium (a mixture of nitrogen and argon) excludes hydrogen embrittlement of the nitrided layer, but this changes the energy of ion nitriding.

Nitriding is used in various industries to improve the operational reliability of a wide range of products: cylinder liners, crankshafts and camshafts, spindles of machine tools, stamping and cutting tools, turbine parts, etc. Durability and performance of instruments is largely determined by the surface condition and its properties. One of the most effective ways of hardening the surface is ion-plasma nitriding.

The aim is to create a nitriding surface layer with high hardness, wear resistance, enhanced fatigue strength and corrosion resistance. Nitriding subjected finished products past mechanical and final heat treatment.

Nitriding has several advantages over cementation. Advantages are a higher hardness and wear resistance of the surface layer, maintaining their high properties when heated to 500C, and high corrosion properties. In the nitrided layer are residual compressive stress, which increases fatigue strength. Furthermore, after the nitriding hardening is not required, thus avoiding defects of the accompanying hardening.

Journal of Physics: Conference Series **567** (2014) 012030 doi:10.1088/1742-6596/567/1/012030

2. Experimental study

The plasma accelerates the reaction, increasing the energy of nitrogen ions, further activating them as a result of cathode sputtering. Plasma nitriding meet environmental requirements, operates a large number of plants for its implementation. A feature of the plasma nitriding is that the process can be controlled much easier compared to the nitriding in a gaseous medium. Requirements to reduce running costs, reduce the temperature and increase the durability of the product during industrial processes become mandatory. In connection with this process, you always get the hardened layer at medium and low temperatures, are widely used in various saturating atmospheres.

At low temperature nitriding steel in various saturating media occurs mainly nitrogen diffusion and the structure and phase composition of the diffusion layer are determined phase diagram of iron - nitrogen. When the high-temperature nitriding steel (> 600 °C) as a saturating medium used only nitrogen, and ammonia.

A method of forming a wear-resistant coating on the surface of articles made of structural steel includes ion nitriding in a plasma environment of the reactive gas - nitrogen surface cleaning components and nitriding. Cleaning of the surface nitriding is carried out at a pressure of reactive gas 5 • 10-3-2 • 10-2 mm Hg. v., the negative bias voltage V and 300-1000 parts of ion current density 2 - 8 mA / cm2 for 30-90 min, cleaning is performed in a plasma of inert gas - argon at a pressure of 3 • 10-4 - 7 • 10-4 mm Hg and a current density of 3 - 5 mA / cm2. The method allows to intensify the process and increase the operational durability of dies experiencing high specific loads during friction.

When nitriding the surface layer formed of compression stress, which increases fatigue strength of components.

Structure and properties of the hardened layer azotiruemyh details depend on the following technical factors: the voltage between the electrodes, the composition of the gaseous medium, degree of dilution, operating temperature, duration of the process, the mutual arrangement of parts and electrodes. Azotiruemye workpieces are mounted in a chamber connected to the negative electrode chamber is sealed and evacuated to a pressure of 1 mm Hg. Art. After evacuation of the air chamber is purged with a working gas for 5-15 minutes at a pressure of 10 mm Hg, then evacuated chamber to a pressure of 0.2 - 0.4 mm Hg, the voltage applied to the electrodes and excite glow discharge. When the voltage 1100-1400 V this stage, the cathode sputtering.

Nitriding temperature is usually 470-580 ° C, strain 400-1100 V, vacuum of 1-10 mm Hg. Working pressure is limited by the properties of the glow discharge. At a pressure below 1 mm Hg ion energy sufficient to heat the workpiece to a working temperature at a pressure above 10 mm Hg. violate the stability of the discharge, glow discharge moves in an arc, which is accompanied by the appearance on the surface of the molten craters.

As nitrogen gas is used, ammonia, nitrogen and a mixture of nitrogen and hydrogen. In the presence of oxygen nitrogen plasma is unacceptable, since it reduces the activity of the working environment; hydrogen has little effect on the growth of the layer. Changes in the concentration of hydrogen in nitrogen-hydrogen mixture in the range of from 1:9 to 9:1 does not affect the parameters of the diffusion layer. Ion nitriding can be carried out in a hydrogen-free plasma. To control the composition of the nitride zone on carbon in the working atmosphere are introduced carbonaceous gases.

Ion nitriding process is realized in two steps: 1) cleaning the surface of cathode sputtering; 2) saturation.

Electrical and vacuum discharge parameters during the cathodic sputtering ensure activation and destruction of a surface oxide film, and a second step to provide surface heating temperature diffusion activity of the gas phase to maintain a concentration gradient of nitrogen on the surface and adjustment process on the phase composition of the diffusion layer.

Cathode sputtering is carried out for 5-60 minutes at a voltage of 1100-1400 V and a pressure of 0.1-0.2 mm Hg. During cathodic sputtering surface temperature of items exceeding 250 °C. The operating parameters of the process upon saturation: U = 400- 100 V, the pressure of 1-10 mm Hg.

Journal of Physics: Conference Series **567** (2014) 012030 doi:10.1088/1742-6596/567/1/012030

After surface treatment for 5-60 min regime cathode sputtering voltage lowered to the working and the pressure was increased to 10.1 mm Hg. When the pressure decreases the length of the cathode glow, which is uniformly distributed over the surface of the part, repeating its outlines. Cathode sputtering may be carried out in a hydrogen atmosphere, the use of which is particularly advantageous during nitriding austenitic steels. Use of a hydrogen plasma nitriding process prior to feeding the ammonia accompanied by decarbonization surface (especially intensively flowing along the grain boundaries) that when a slight fall improves ductility hardness of the diffusion layer.

Operating process temperature (470-580 ° C) is achieved in 15-30 minutes. The heating rate is determined by the ratio of the surface and the mass of details.

After isothermal holding parts are cooled to room temperature under vacuum. In this cooling rate as compared with the above heating furnace, since the ion treatment is only heated product, and heating the walls of the container (working chamber) due to heat radiation and convection under vacuum negligible.

Equipment for ion nitriding and regulation should provide for stable maintenance of electrical and vacuum discharge characteristics. Installation for nitriding in a glow discharge consists of four functional systems: electrical, vacuum, gas distribution and automatic control of the process parameters. The control system includes a digital controller for controlling the entire operation of the unit and provides the following controls:

- Pressure in the chamber;
- The temperature of the parts;
- Heating rate;
- The composition of the gases.

The control system consists of a master controller and slave nodes (transmitters and drivers) connected through a highly reliable communication network. The controller receives data from the transmitters, the software executes in real time and sends commands to drivers who drive machinery installation.

Block software controls the data collection, the work pressure regulators, current and temperature, logic control, data logging, control of the technological process, error detection and visualization. Long-term memory of the controller 50 to store the history of technological regimes and technological processes. In the history includes the most important parameters - temperature, pressure, current, voltage, gas flow and time. Software running on the system Windows98 / XP, the PC communicates with the main controller. Software display the current setup parameters to the computer display, maintains a history of technological processes in the database and is reviewing the data in graphical and tabular form.

3. Results of experiment

The plasma generator excites the pulse glow electrical discharge of the following parameters:

Voltage 500 - 800 V

The maximum pulse current of 25 - 250 A

The pulse rate (adjustable) 5 - 10 kHz

The fill factor (adjustable) 50 - 80%

The plasma generator and equipment supply and protection of: - electrical filters, circuit breakers, contactors, power transformers and power thyristor rectifier - housed in a metal cabinet IP-54 air cooling.

4. Conclusions

Changes in current density in the high range (0.5-20 mA / cm2) does not affect the nitriding process.

Improving technical efficiency and technological complex is to optimize the process parameters, depending on the dimensions and weight and geometric details. It is because of these characteristics depends on the uniformity of the density distribution of the ionized gas on the surface of the component, since the intensity of the glow discharge plasma in the nitrogen atmosphere is increased by the sharp edges of parts.

References

[1] Arzamasov BN, Bratukhin AG, Eliseev YS, Panayiotou TA. *Ionic chemical heat treatment of alloys in gas media* (Moscow: Izd.MGTU them. NE Bauman, 1999) p 400
[2] Liakhovich LS. *Case hardening alloys* (Moscow .: Metallurgy, 1981) p 420
[3] Lahtin YM, Arzamas BN. *Case hardening* (Moscow.: Metallurgy, 1985) 256 p.
[4] VG Kaplun, Shepherd IM. *Plasma diffusion processes of surface modification of materials, technology and equipment* (Proceedings of the scientific-practical Symposium "Equipment itehnologii heat treatment of metals and alloys in engineering" - Kharkiv (Ukraine), OTTOM, -2000) p 145-154.
[5] Israphilov D I 2007 *Management of a plasma technological complex for heat treatment of parts with specified strength characteristics* (Naberezhnye Chelny: Kama state engineering and economic academy) p 141

Journal of Physics: Conference Series **567** (2014) 012031 doi:10.1088/1742-6596/567/1/012031

Control of glow discharge parameters using transverse supersonic gas flow - numerical experiment

A I Saifutdinov[1,2], B A Timerkaev[3] and B R Zalyaliev[3]
[1]Kazan Federal University, 18 Kremlyovskaya str., Kazan 420008, Russian Federation
[2]Saint Petersburg State University, S.Petersburg, Russian Federation
[2]Kazan National Research Technical University after name A.N. Tupolev – KAI

E-mail: as.uav@bk.ru, btimerkaev@gmail.com

Abstract. A low pressure glow discharge in a transverse supersonic gas flow was studied by numerical modelling for the case where the flow only partially fills the interelectrode gap. It's shown that by organizing a supersonic gas flow in a limited region of the interelectrode space can be controlled combustion conditions of the glow discharge, and its parameters. It is shown that it is possible to achieve stable combustion glow discharge at low and superlow pressures, when the parameter *pL* lies on the left branch of the Paschen curve.

1. Introduction

The control of the parameters of the DC glow discharge's plasma is of considerable fundamental and applied interest. Glow discharges at low pressures and long discharge gaps are promising for the application of coatings, plasma surface modification and others.

As is known [1, et al.], one of the necessary conditions for the existence of glow discharge is the presence of all near cathode regions. The length of these regions increases with decreasing pressure. This is due to the fact that the length of each zone is mainly determined by the number of ionizing collisions of electrons with neutral gas particles. As a result, under high vacuum for self-organization glow discharge may be not enough distance between the electrodes. In this case, the discharge becomes difficult that is combustion conditions correspond to a point belonging to the left branch of the Paschen curve [1], or does not lights up.

To solve this problem, the article [2] proposed a method which consists in the organization of a supersonic gas flow in the direction of perpendicular to the electric field in the area next to the cathode layer CS. In this case, on the way of the electrons, which gain energy in CS, in the narrow part of the interelectrode space in which is organized transverse supersonic flow, by artificial means created "the stockade" of neutral gas particles, i.e., to create conditions similar increase in the concentration of neutral particles, which leads to increase the number of ionizing collisions. As a result, all main regions of glow discharge are formed, including the positive column and PC.

It is worth noting works [3-8], which been studied in detail glow discharge in longitudinal and transverse supersonic gas flow.

The aim of this work is to demonstrate by using of numerical experiments in the approach of hybrid model the new method for controlling parameters of glow discharge by using transverse supersonic gas flow.

2.The model description

One of the most common methods of modeling the spatial structure of the glow discharge is diffusion-drift (fluid) approximation [1, 3, 6, and others]. However, at low pressures, when clearly observed the near cathode regions of the discharge, in particular, the region of the negative glow NG and the Faraday dark space FDS must be considered non-locality of the ionization processes occurring in these areas. Different approaches to describe the near cathode regions of the discharge have been proposed earlier in papers [7, 8-11].

In this paper, for the numerical experiments are formulated hybrid model [9], glow discharge, based on the balance equations for the concentrations of electrons n_e, positive n_+ and negative ions n_-, heat balance equation of electrons n_ε, which takes into account not only the bulk processes, but also the spatial transfer by conduction and Poisson equation for finding the self-consistent electric potential φ. The calculations were performed for air (77% N_2, 23% O_2) with a set of plasma-chemical reactions described in [14-15]. In the energy balance equation for the electrons were taken into account members of the excitation and elastic losses.

The system of equations describing the structure of the glow discharge, is written as follows:

$$\frac{\partial n_k}{\partial t}+\nabla\cdot\mathbf{\Gamma}_k=S_k\,,\ \mathbf{\Gamma}_k=-D_k\nabla n_k+z_k\mu_k\mathbf{E}n_k\,, \tag{1}$$

$$S_e=(\nu_{ion}-\nu_{att})n_e+\nu_d n_- -\beta_{e+}n_e n_+\,,\quad S_-=\nu_{att}n_e+\nu_d n_- -\beta_{ii}n_- n_+\,, \tag{2}$$

$$\frac{3}{2}\frac{\partial n_\varepsilon}{\partial t}+\nabla\cdot\mathbf{Q}_\varepsilon=-e\mathbf{\Gamma}_e\cdot\mathbf{E}+S_{el}+S_{in}\,,\quad \mathbf{Q}_\varepsilon=-D_\varepsilon\nabla n_\varepsilon-\mu_\varepsilon\mathbf{E}n_\varepsilon\,,\quad Q_+=Q_e+Q_-\,, \tag{3}$$

$$\frac{\partial(\rho T)}{\partial t}-\nabla\chi\nabla T=e\mathbf{\Gamma}\cdot\mathbf{E}\,, \tag{4}$$

$$\Delta\varphi=-\frac{q_e}{\varepsilon_0}\left(n_+-n_- -n_e\right). \tag{5}$$

where $\mathbf{\Gamma}_k$ – are fluxes of electrons, positive and negative ions, $\mathbf{Q}_\varepsilon$ – are the energy flux of the electrons; z_k – is the electric charge of the particle; D and μ with index «k» – are the diffusion coefficients and the mobility of electrons, positive and negative ions, respectively; with index « ε » – are the diffusion coefficient and the mobility of electron energy; $\mathbf{E}=-\nabla\varphi$ – is the electric field strength, ν_{ion}, ν_{att}, ν_d – are the ionization, attachment and detachment frequencies; β_{e+}, β_{ii} – are the coefficients for electron-ion and ion-ion recombination. The ionization frequency calculated by convolution with a Maxwellian distribution function with section $\sigma(w)$ of the corresponding process [13].

The first term on the right hand side of the first expression (3) corresponds to the Joule heating of the electrons in an electric field. The second term S_{el} describes the energy loss due to elastic collisions of electrons with atoms and molecules. The third term S_{in} describes the inelastic energy loss of an electron gas.

Mobility is determined as follows:

$$\mu_e=q_e/m_e\nu_{el}N_0,\quad \mu_+=\left(1.7\cdot10^{-4}/p\right)m^2/(V\cdot s),\quad \mu_\varepsilon=(5/3)\mu_e, \tag{6}$$

where ν_{el} – is frequency of elastic electron-atom strikes, p – pressure (Torr). The diffusion coefficients were calculated by means of the Einstein relation:

$$D_k = T\mu_k,\ D_e = T_e\mu_e,\ D_\varepsilon = T\mu_\varepsilon. \quad (7)$$

Boundary conditions were set at the cathode and anode for the electron density, the energy density of the electrons and the ion concentration and the electric potential

$$\mathbf{n}\cdot\mathbf{\Gamma}_{\mathbf{e}}\big|_0 = \frac{\nu_{th\,e} n_e}{2} - \gamma\,\mathbf{n}\cdot\mathbf{\Gamma}_k,\quad \mathbf{n}\cdot\mathbf{\Gamma}_{\mathbf{e}}\big|_L = \frac{\nu_{th\,e} n_e}{2}, \quad (8)$$

$$\mathbf{n}\cdot\mathbf{Q}\big|_0 = \frac{5}{6}\nu_{th\,e} n_e,\quad \mathbf{n}\cdot\mathbf{Q}\big|_L = \frac{5}{6}\nu_{th\,e} n_e, \quad (9)$$

$$\mathbf{n}\cdot\mathbf{\Gamma}_k\big|_0 = \nu_{i\,th} n_k,\quad \mathbf{n}\cdot\mathbf{\Gamma}_k\big|_L = \nu_{i\,th} n_k + \mu_k\,\mathbf{n}\cdot\mathbf{E}, \quad (10)$$

$$\varphi(0,r)=0,\ \varphi(L,r)=U_0. \quad (11)$$

где $\gamma = 0.1$ – is the coefficient of secondary electron emission from the cathode, $\nu_{e\,th} = \sqrt{8k_BT_e/\pi m_e}$ и $\nu_{i\,th} = \sqrt{8k_BT_e/\pi M}$ – are the average thermal velocity of the electrons and the atoms of the gas, respectively, (T – in eV).

3. The results of numerical experiments.

Equations (1)-(5) with boundary conditions (8)-(11) were solved for a one-dimensional geometry with the distance between the electrodes L = 4 cm. Picture of the distribution of the plasma parameters of a glow discharge at a pressure $p = 0.15\,Torr$ and the voltage between the electrodes U = 500V is shown in Fig. 1. Clearly observed the near cathode region of the discharge – cathode layer CS, a region of negative glow NG and the Faraday dark space FDS. The positive column PC is missing. Under these conditions, the discharge parameter $pL = 6.0\,Torr\cdot cm$ belongs to the minimum of the Paschen curve for air [1].

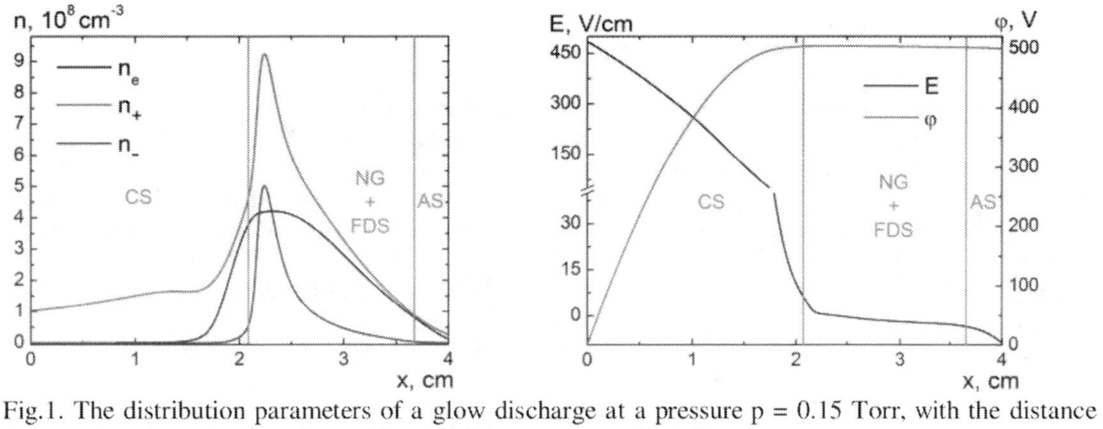

Fig.1. The distribution parameters of a glow discharge at a pressure p = 0.15 Torr, with the distance between the electrodes L = 4cm without making a supersonic airflow.

Through supersonic gas flow perpendicular to the direction of the electric field in a limited region of the interelectrode space, and varying the gas flow rate $G = \rho V_{super} S$, which is related to the density of the supersonic flow ρ and its speed V_{super} (assuming that the gas density and flow rate does not vary across the cross section) can thereby increase the concentration of neutral particles in the selected area of the discharge gap is several times. Furthermore, when implementing the supersonic gas flow possible to light discharge in conditions when the parameter *pL* shifted to the left side of the Paschen curve for the same voltage between the electrodes.

Thus in Fig. 3 shows the structure of a glow discharge at a pressure $p = 0.1 Torr$ in the discharge chamber and the implementation of a supersonic gas flow region $x \in (0.25 - 0.35)\, cm$. It is worth noting that the implementation of the supersonic gas flow possible to light the discharge in conditions when the parameter pL displaced in the left branch of the Paschen curve at the same voltage between the electrodes. It is seen that NG and FDS narrowed and focusing entirely in the area with a supersonic flow. In addition, the formed PC, a small part of which is localized in the region with a supersonic flowand the remainder (the majority) - after area with a supersonic flow. Similar results are observed in laboratory experiments [2].

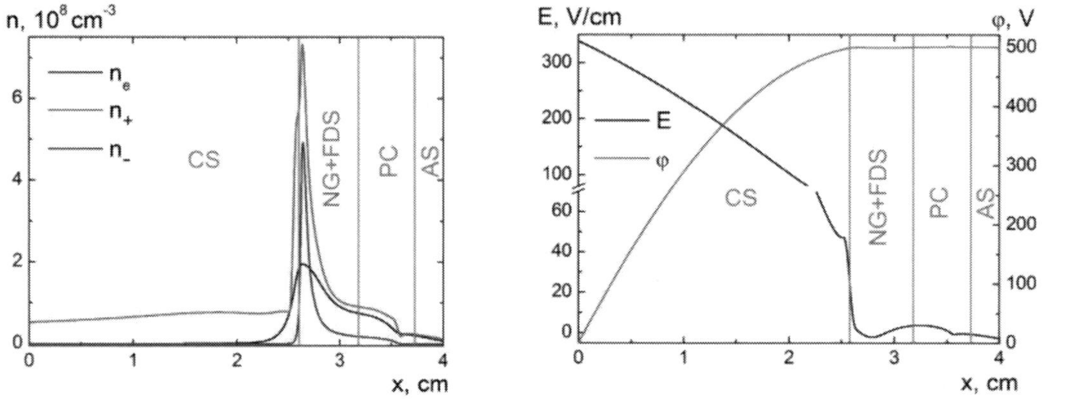

Fig.2. The distribution parameters of the discharge at a pressure of p = 0.1 Torr considering supersonic gas flow in the region from 2.5 cm to 3.5 cm with a gas consumption corresponding increased concentrations of neutral particles 10 times.

4. Conclusions

Thus, it is shown that organizing a supersonic gas flow in a limited region of the interelectrode space can be controlled as a glow discharge conditions, and its parameters. In particular, it is shown that it is possible to achieve stable combustion glow discharge at low and superlow pressures, when the the parameter pL lies on the left branch of the Paschen curve. In this case, there is an extended cathode layer and the plasma region - negative glow NG, Faraday dark space and positive column FDS PC concentrated in a limited region with a supersonic gas flow. The obtained results allow to conclude that found a new effective way to control the longitudinal structure of the discharge, which can have a variety of applications, both in the region of the plasma coating, and in gas lasers.

The work is performed according to the Russian Government Program of Competitive Growth of Kazan Federal University.

References

[1] Raiser Y P. *Physics of gas discharge*. M .: Nauka, Fizmatlit, 1987, 1992; Publisher Intellect, Dolgoprudnyi 2009.

[2] Timerkaev B A, Zalyaliev B R, *A glow discharge in a transverse supersonic gas flow at low pressures* 2014 High Temperature **52** 489-493.

[3] Dautov G , Timerkaev B, *Generators nonequilibrium plasma*. Feng, Kazan, 1996.

[4] Galeev I, Gonchar, Timerkaev B 1990 High Temperature **28** p 843
[5] Galeev I, Gonchar, Timerkaev B 1992 High Temperature. **30** p 439
[6] Surzhikov S T, *Physical mechanics of gas discharge*, MSTU. NE Bauman, Moscow (2006).
[7] Aleksandrov A L, Schweigert I V, *Interaction of surface glow discharge with a gas flow* 2010 JETP **137**, 966-972.
[8] Boef J P 1987 Phys. Rev. A **36** 2782
[9] Boef J P, Pitchford L C 1995 Phys. Rev. E **51** 1376 .
[10] Rafatov I, Bogdanov E A and Kudryavtsev A A *Account of nonlocal ionization by fast electrons in the fluid models of a direct current glow discharge* 2012 Phys. Plasmas **19** 093503
[11] Saifutdinov A I, Timerkaev B A *Drift model of a glow discharge with a non-local values of the electric field in the ionization source* 2012 Journal of Engineering Physics **85** 1202-1207.
[13] See Compilation of atomic and molecular data, assembled and evaluated by AV Phelps: ftp://jila.colorado.edu/collision data / eletrans.txt
[14] Kossyi I A, Kostinsky A Yu, Matveyev A A and Silakov V P *Kinetic scheme of the non-equilibrium discharge in nitrogen-oxygen mixtures* 1992 Plasma Sources Sci. Technol. **1** 207-220
[15] Macheret S O, Shneider M N, Miles R B, *Modeling of air plasma generation by repetitive high-voltage nanosecond pulses* 2002 IEEE Trans. Plasma Sci. **30** 1301-1314

Glow Discharge Characteristics in Transverse Supersonic Air Flow

B A Timerkaev[1], B R Zalyaliev[1], A I Saifutdinov[2,3]
[1]Kazan National Research Technical University after name A.N. Tupolev – KAI
[2]Kazan Federal University, 18 Kremlyovskaya str., Kazan 420008, Russian Federation
[2]Saint Petersburg State University, S.Petersburg, Russian Federation

E-mail: btimerkaev@gmail.com, zalyaliev@bk.ru, as.uav@bk.ru

Abstract. A low pressure glow discharge in a transverse supersonic gas flow of air at pressures of the order 1 torr has been experimentally studied for the case where the flow only partially fills the inter electrode gap. It is shown that the space region with supersonic gas flow has a higher concentration of gas particles and, therefore, works as a charged particle generator. The near electrode regions of glow discharge are concentrated specifically in this region. This structure of glow discharge is promising for plasma deposition of coatings under ultralow pressures

1. Introduction

One of the necessary conditions for the existence of glow discharge is the presence of all near electrode regions. The length of these regions increases with a decrease in pressure. This is due to the fact that the length of each zone is mainly determined by the number of ionizing collisions of electrons with neutral particles. This result leads to the fact that in a high vacuum for self-organization it lacks the interelectrode space, and the discharge fails to ignite.

In devices of the plasma coating, this problem is solved by applying a magnetic field, which significantly increases the possibility of collisions between electrons and neutral particles due to the combined action of electric and magnetic fields on moving charges and provides the necessary conditions for the existence of the discharge.

However, there is another approach, which implies formation of different concentrations of neutral atoms in different regions of the interelectrode gap. The concentration of gas particles near the cathode should be such in order to mean free path was greater than the distance between the electrodes and ejected from the cathode due to ion bombardment, were not scattered by neutral gas particles. Whereas the concentration near the anode should be sufficient for the electron to undergo several tens of collisions. These requirements can be satisfied by providing supersonic gas circulation in the direction perpendicular to the electric field.

The glow discharge in a supersonic gas flow in supersonic nozzles with a central body was studied in [1–5], where highlighted advantages of supersonic gas flow through the discharge region. In these studies, the discharge was ignited between the nozzle-anode and cathode-central body. In such devices, the glow discharge was self-organization thus in order to for the discharge was needed minimum voltage. In these studies, was noted the stability conditions even at high power discharge, because the thermal energy in the discharge effectively took away gas flow. Also the influence of supersonic flow on the discharge was considered in [5].

In paper [6] investigated the glow discharge at low pressures with different density of gas in the discharge gap. Unequal particle density in the discharge gap created by the system diffuser - confusor which provides an increased concentration of the gas particles in the area, the boundaries of which are the boundaries of the supersonic flow

2.Experiment

Block diagram of the experimental setup for investigating glow discharge in a transverse supersonic gas flow is shown in Fig. 1. The setup is a vacuum chamber with two platelike copper electrodes placed opposite to each other at a distance of 5 cm. Pressure in the chamber was 1 Torr. Supersonic pumping system carried through a Laval nozzle - confusor. Confusor connected to a vacuum pump, and through it occurred pumping gas from the chamber.

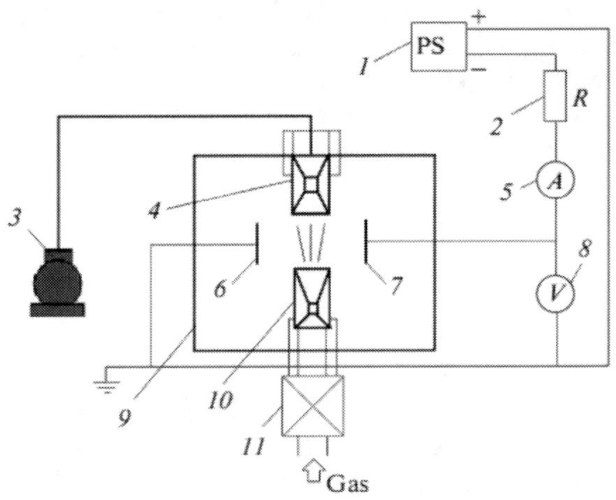

Fig. 1. Schematic of the experimental setup: (1) power supply, (2) ballast resistance, (3) vacuum pump, (4) confusor, (5) amperemeter, (6) anode, (7) cathode, (8) voltmeter, (9) vacuum chamber, (10) Laval nozzle, and (11) gas inlet valve.

Due to low pressure in the chamber al the entire interelectrode gap consists of near cathode regions that are necessary for maintaining discharge, also observed negative glow, the cathode and anode spot (Fig. 2.1, 2.2). Feed atmospheric air through a Laval nozzle in the region between the anode and cathode leads to the appearance the positive column. Discharge distinctly shows the boundaries of the gas flow (ris.3.1-3.4).

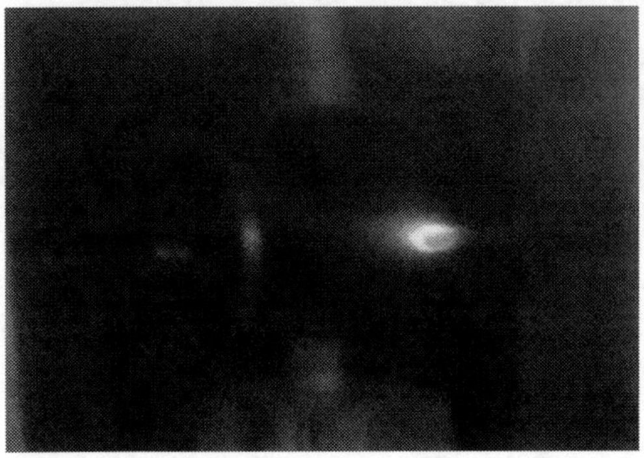

Fig. 2.1. Glow discharge at copper electrodes in

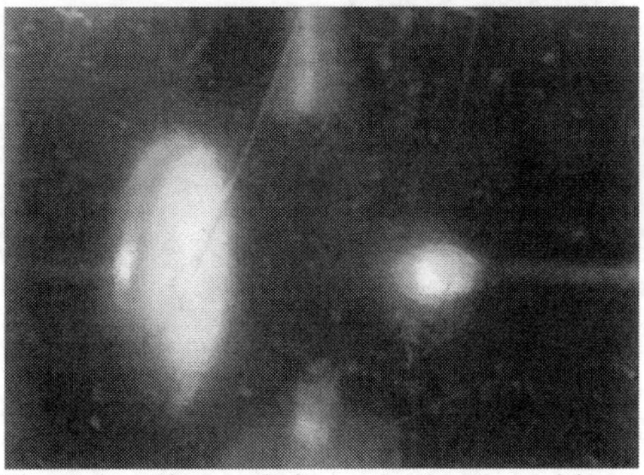

Fig. 2.2. Photo of the cathode spot without

air under low pressure without supersonic flow. Cathode left, anode right.

supersonic flow. Cathode left, anode right.

The boundary of the supersonic flow coincides with the boundary of the positive column and the Faraday dark space. Due to reduce of the mean free path in a supersonic gas flow is declining cathode region. At that the gas pressure in the chamber is not changed, since the gas flow hits the confusor and successfully removed from the discharge area. Positive column has sharp boundaries. Carried out a series of experiments to study the effect of a supersonic gas flow on the location of the positive column. In these experiments, a supersonic flow was organized in different parts of the electrode gap. In all these experiments, the observed binding of the positive column border to the border region of supersonic flow.

Also, when certain expenses of air supplied to the discharge region, there was a lack of any glow only in the discharge gap, the boundaries of which coincide with the boundaries of the flow (Figure 3.2), as if the flow of "cuts" discharge. This can be explained by the fact that in the area of influence of flow on the discharge sharply increases the concentration of neutral atoms. Electrons, which committed acts of ionization has is not enough energy to excite the atoms of the air in the segment flow restriction of in this area, and there is no glow, which observed after border flow, since electrons can gain energy.

Along with the reduction of the cathode region and the emergence of the positive column with a clear boundary, and the effect of "cutting" the discharge was discovered phenomenon brightly glowing border outbound flow from the nozzle (Figure 3.3), and at a certain flow rate observed two boundaries, between which there is no illumination (Fig. 3.4).

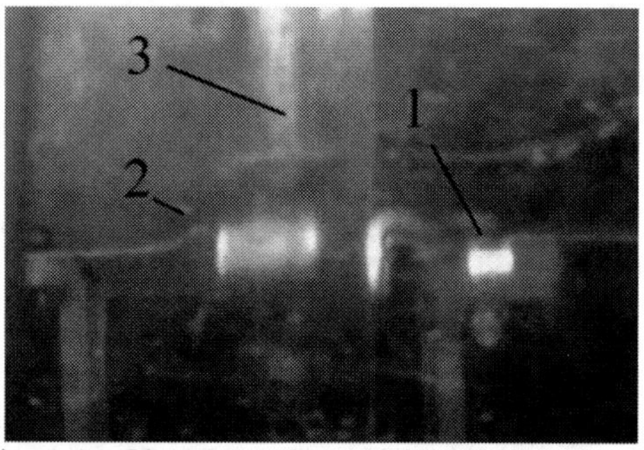

Fig. 3.1. Clear boundary of the positive column in the presence supersonic flow. 1 – cathode 2 – anode 3 – confusor.

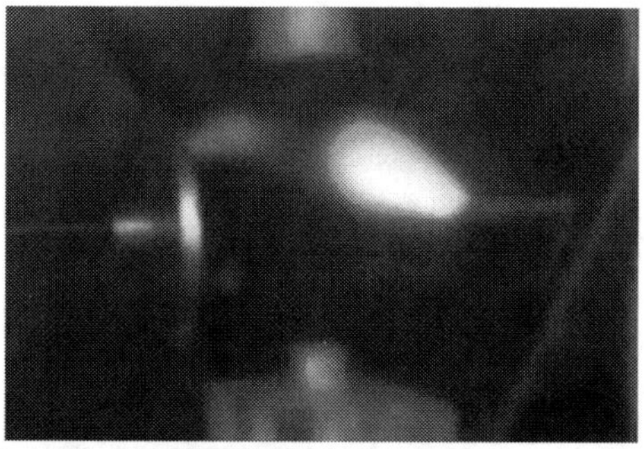

Fig. 3.2. "Cutting" discharge supersonic flow. Cathode left, anode right.

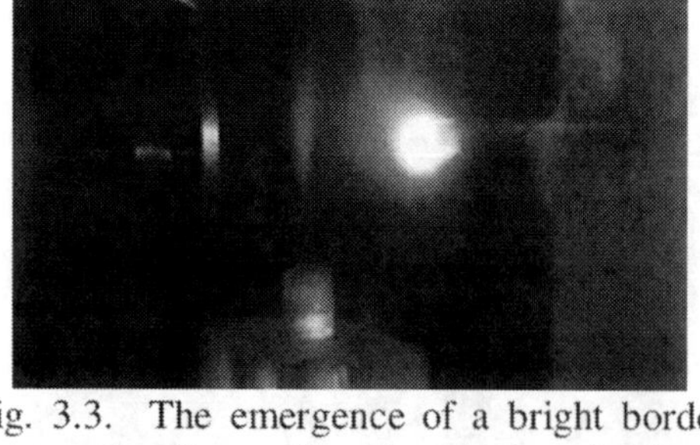

Fig. 3.3. The emergence of a bright border supersonic flow after the supersonic flow. Cathode left, anode right.

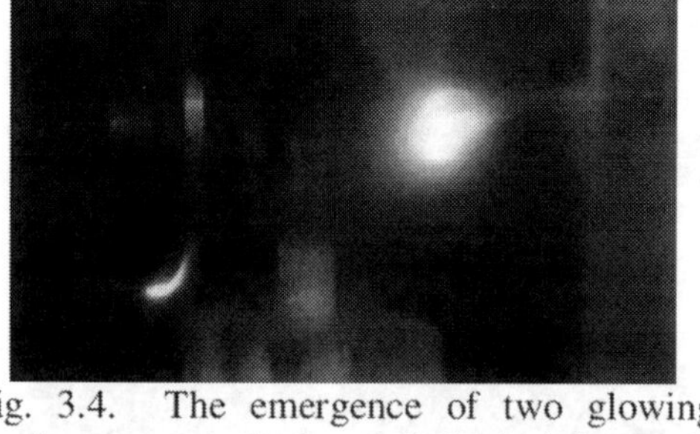

Fig. 3.4. The emergence of two glowing borders supersonic flow. Cathode left, anode right.

Moving away from the cathode, the electrons are gradually gaining sufficient energy to excite atoms, approaching the first boundary of the stream, where the concentration of neutral particles is already significantly higher. Therefore we see the first glowing flow boundary. Further, electrons pass into area where the reduced mean free path and they will not have enough energy to excite the atoms of air.

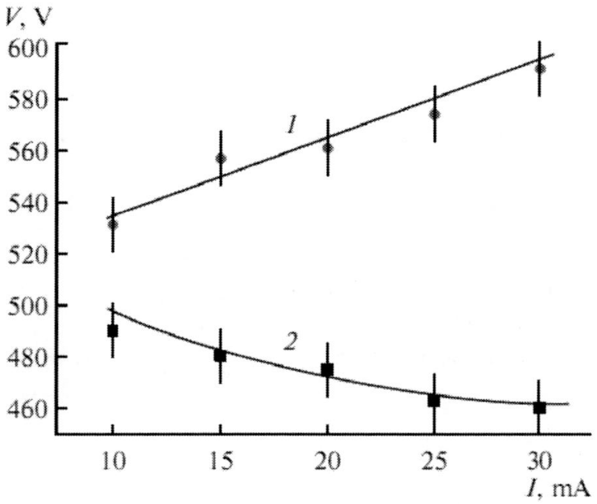

Fig. 4. I–V characteristic of the glow discharge (1) in supersonic flow and (2) in its absence.

Figure 4 (curve 1) shows the I–V characteristic of the discharge with the supersonic flow switched on. It can be seen that this characteristic began rising and the discharge voltage is higher than that in the absence of gas circulation. The increase in the discharge voltage can easily be explained by the formation of an extended positive column. The voltage drop at the positive column additionally increases the discharge voltage. However, the increase in the discharge voltage depends on the position of the supersonic flow in the interelectrode gap. For example, when the supersonic flow is formed near the cathode, the discharge voltage increases even more. The fact that the I–V characteristic begins rising after the formation of the supersonic flow can be explained by the influence of the flow on the characteristics of both the positive column and the near electrode regions.

4. Conclusions

The formation of a supersonic gas flow in the limited region of the interelectrode gap makes it possible to control the near electrode processes in a glow discharge. The results obtained determine a new tendency in the technology of plasma deposition of coatings. Transverse supersonic gas circulation allows one to use plasma deposition systems without constant magnets and to increase significantly the cathode sputtering efficiency.

The work is performed according to the Russian Government Program of Competitive Growth of Kazan Federal University.

References

[1] Dautov G , Timerkaev B, *Generators nonequilibrium plasma*. Feng, Kazan, 1996.
[2] Galeev I, Gonchar, Timerkaev B 1990 High Temperature **28** p 843
[3] Galeev I, Gonchar, Timerkaev B 1992 High Temperature. **30** p 439
[4] Alferov V I Bushmin A S and Dmitriev A M 1985 Prib.Tekh. Eksp. **1** p 143
[5] Efimov B G Ivanov V V etc. 2011 High Temperature **49** p 479
[6] Timerkaev B A, Zalyaliev B R *A glow discharge in a transverse supersonic gas flow at low pressures* 2014 High Temperature **52** 489-493.

The acoustic model of oscillations of gas combustion in coaxial pipes

E. V. Semenova[1], V. M. Larionov[1], E. I. Kazakova[1]

[1]Kazan (Volga region) federal university

E-mail: evgeniya.yallina@gmail.com

Abstract. Organization of pulse combustion mode is one of the possible solutions to the problem of energy efficiency installations using hydrocarbon fuel. For grate combustion of solid fuels, in particular, solid industrial wastes are considered to be promising coaxial system, allowing the admission of secondary air to the combustion zone. In this paper we proposed an acoustic model of oscillations of gas when burning solid fuel in the system is coaxially arranged pipes with natural air supply. The description of the motion of the gas in the system during one period of oscillation.

1. Introduction

It is known that when pulsating combustion is increasing combustion efficiency and acceleration of heat and mass transfer processes in comparison with a uniform flame. These advantages can be used to solve problems and energy saving and the environment.

The aim of this work was to investigate the acoustic properties of the system, obtaining ratios, determining the conditions for the excitation frequency and amplitude of oscillations of the gas

2. Calculations and physical description of the fluctuation of gas

In [1] was developed a mathematical model of the vibration excitation gas when burning solid fuels in coaxial pipes (Fig.1).

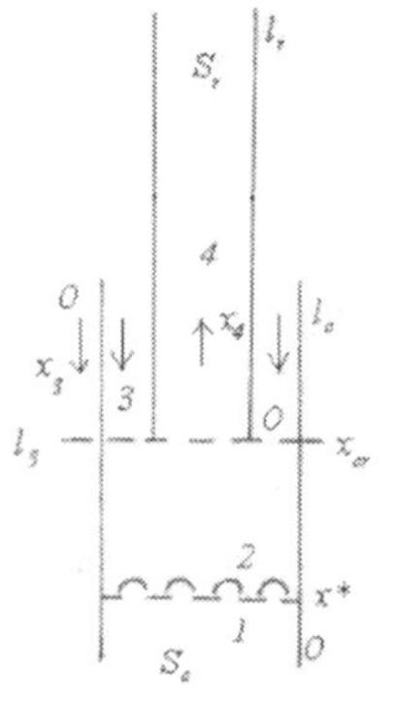

Fig. 1. The model of a coaxial combustor

Journal of Physics: Conference Series **567** (2014) 012033 doi:10.1088/1742-6596/567/1/012033

Scroll to the study for the installation of four sections. The first part of the combustion chamber to the grid that holds the fuel. The second area is the area of heat dissipation, which is located between the grille and the cross-section of the combustion chamber, which introduced a resonance tube. The third section is an annular channel between the tubes, which served the secondary air. The fourth plot is a resonance tube, which moves the hot gas.

The calculations were performed by power method [2,3] for the combustion chamber with an inner diameter of 60 mm, a resonance tube with an inner diameter of 36 mm and a wall thickness of 1 mm, a gas Temperature of 1073 K in the combustion zone and a gas temperature of 673 K at the output of the resonance tube length 1.2 m is taken from [1]. The analysis of oscillations of gas was carried out under the condition that the length of the combustion chamber is 0.4 m, the length of the resonant pipe is 1.2 m, $x^* = 0.24$ m, $x_{c,r} = x^* + 0.01\text{m}$ In this case, the excitation of the gas oscillations with a frequency of 201 Hz and a sound pressure level in the combustion zone 136,7 dB.

For greater clarity, using a single for all sections of the reference system, and as one of the coordinates of the selected longitudinal coordinate, corresponding to the combustion chamber. Replacement of coordinates is performed as follows $x_3 = l_c - x,\ x_4 = x - x_{c,r}$.

As a single parameter that defines the amplitude of the pulsation velocity and pressure in the system will take the amplitude of the pressure pulsations in the combustion zone[1]:

$$p_* = p_{1,m}\left|\sin\left(\frac{\omega x^*}{c_1} + \varphi_1\right)\right|,$$

where p^* is the amplitude of the pressure fluctuations in the cross-section $x = x^*$, $p_{1,m}$ is the amplitude of the pressure fluctuations of gas in the combustion chamber, ω - angular (circular) frequency c_1=343 m/s is the speed of sound, φ_1 =0,61 ωRc/c_1 , R_c is the radius of the combustion chamber.

Substituting this expression into the equations describing the fluctuations of velocity and pressure in the system [1], we can calculate the magnitude of the pulsations of velocity and pressure at any cross-section of each section of the system at any point in time.

It was assumed that at the initial moment of time the pulsations of the gas pressure at all points of the system is zero. At the same time, the amplitude of the pulsations of the gas velocity in each point of the system has a maximum value (Fig. 2, a). In the lower part of the combustion chamber (graph 1) pulse speed primary air is directed from the inlet to the combustion zone. In the annular channel pulsatile velocity secondary air though and has the opposite direction, but also as in the first case directed towards the combustion zone. Therefore, at time t = 0 in the combustion zone receives the maximum amount of air. The amount of the pulsation flow volumes of the primary and secondary air is equal to the pulsation of the volume flow of combustion products included in the resonance tube (Fig. 3, a).

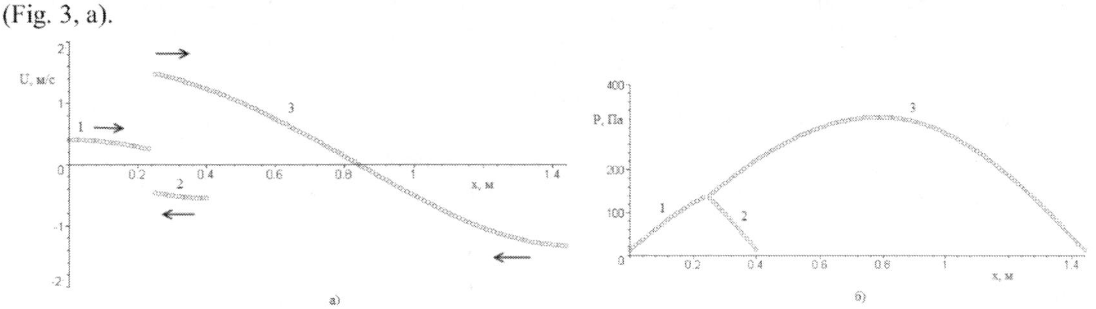

Fig. 2. Distribution of (a) the velocity uctuation of the gas in the system ; b) pressure pulsations in the system when $t = T/4$

Therefore, the pulsation velocity of the gas in each part of the resonance tube is directed away from the entrance to the middle (left part of the graph 3). In section x≈0.64 m, located somewhat below the middle of the resonance tube, the pulsation rate of the gas is zero. A characteristic feature is that the amplitude of the pulsation velocity inlet gas resonance tube is less than the open exit. For comparison, we note that the distribution of the gas velocity in the pipe, open at the ends, symmetrically relative to the middle.

Despite the equality of the pulsation flow volumes of cold air and hot combustion products, the difference in density leads to an increase in the mass of the gas in the combustion zone (Fig. 3, b). Therefore, over time the pressure of the gas in the combustion zone begins to rise. This reduces the pulsation rate of the primary and secondary air. The increase in the mass of gas in the middle part of the resonance pipe increases the gas pressure in this area and causes a reduction of the pulsation of the gas velocity across the pipe. After a quarter of the oscillation period pulsation velocity of the gas in the entire system becomes zero. As expected, at the entrance of the combustion chamber, the upper end of the annular channel and the output of the resonant tubes are open, the pressure increases slightly (Fig. 2, b). When approaching the combustion zone, the air pressure in the combustion chamber and the annular channel increases and reaches a value equal to the pressure of the hot gas inlet in the resonance tube. However, the greatest compression of the hot gas is in section x≈0,64 m, where, as previously established, there is no pulsation of the gas velocity.

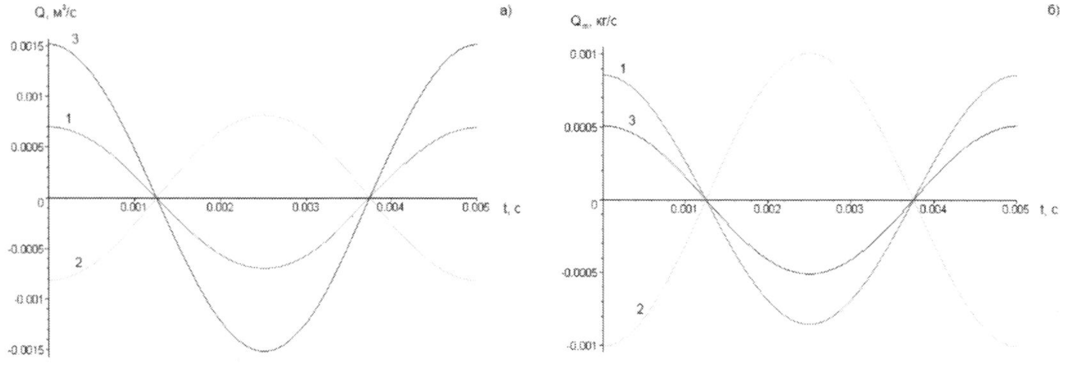

Fig. 3. Changes ripple volumetric gas flow (a) and pulsation mass flow rates of gas (b) in the combustion zone for the period of oscillation of gas: 1 - primary air, 2 - secondary air, 3 - combustion products

The gradient of gas pressure at sites 1 and 3 such that the direction of the velocity uctuation of the air (Fig. 4, graphs 1, 2) becomes the opposite of what it was at t = 0 (Fig. 2, graphs 1, 2). This leads to a reduction of the mass of gas in the combustion zone, which is not compensated by the inflow of gas from the resonance tube (Fig. 3, b, the lower part of the graphs 3). The gas pressure in the combustion zone begins to decrease. In the resonance pipe is gas flow, since the pressure in the middle part of the tube exceeds the pressure at its ends (Fig. 2, b). The pressure in this pipe is gradually reduced. At the time when $t = T/2$ the pressure in the entire system becomes equal to its average value, that is, the sound pressure becomes equal to zero. Pulsatile velocity of the gas over the entire system reaches maximum values and aims according to Fig. 4, and. Despite the fact that the pressure gradient vanishes, the gas particles continue to move by inertia. Reducing the mass of the gas in the combustion zone and in the middle part of the resonance pipe leads to a decrease in gas pressure. At each site of the system is the pressure gradient, which inhibits the movement of gas. At the moment when $t = 3T/4$ there is a maximum negative pressure of gas in the system, and pulsation velocity of the gas becomes equal to zero. In the middle part of the resonance pipe - maximum dilution gas, and the pressure in the cross section x≈0.64 m lower than in the combustion zone (Fig. 4, b, graph 3).

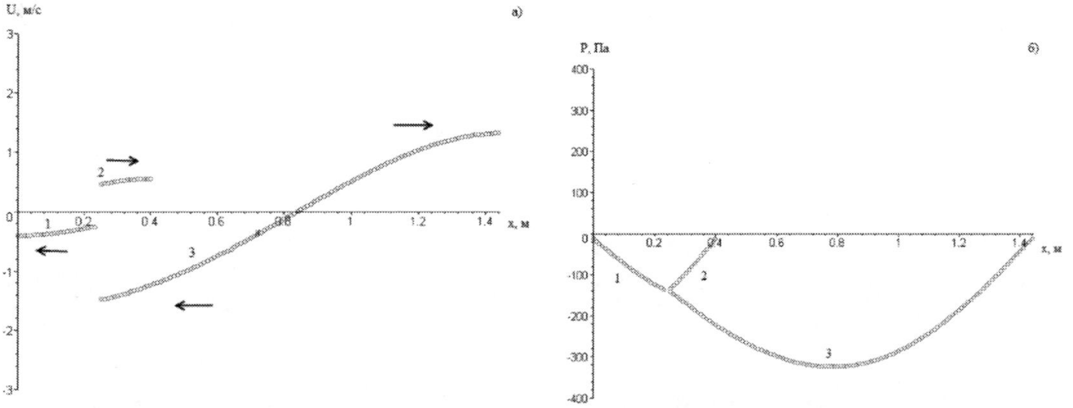

Fig. 4. Distribution of (a) the velocity uctuation in the gas system when $t = T/2$; b)pressure pulsations in the system when $t = 3T/4$

The gradient of gas pressure in the lower part of the combustion chamber, the annular channel, the lower and upper parts of the resonance pipe is such that further leads to pulsating gas movement in the directions indicated by the arrows in Fig. 2, and. The gas is gradually accelerated. Thus there is an increase in the mass of the gas in the combustion zone and the resonant pipe (Fig. 3, b). This leads to a gradual increase of pressure up to its average value in the system. After the period of oscillation of the distribution of the velocity uctuation and the gas pressure in the system take the form as in the initial moment of time (Fig. 2, a).

So, as a result of the calculations is given a vivid physical description of the gas oscillations occurring during pulsation combustion of solid fuel is coaxially arranged pipes.

This work was funded by the subsidy of the Russian Government to support the Program of Competitive Growth of Kazan Federal University among World's Leading Academic Centers

References

[1] Larionov V. M., Yallina E.V., Iovleva O.V. 2014 Problems of energy 2014 **3-4** 3-14
[2] Larionov V. M., Nazarenko T.I. 2000 Izv. VUZ. Aviatsionnaya Tekhnika 2000 **4** 68-69
[3] Iovleva O.V., Larionov V. M. 2007 Izv. VUZ. Aviatsionnaya Tekhnika 2007 **3** 50-53

Synthesis of oxidic powder in nonequilibrium low-temperature plasma with increase of indicators of quality of process

Ruz R Saubanov[1], V V Zvezdin[1],I H Israfilov[1], I I Haybullin[2], R R Rakhimov[1],

[1]Naberezhnye Chelny institute (branch) of KFU, Naberezhnye Chelny, Russia
[2]NaberezhnyeChelny, Russia

E-mail:irmaris@yandex.ru, saubanov@mail.ru

Abstract. Ways of synthesis of powders in low-temperature plasma define various approaches in the solution of technological tasks for receiving difficult oxidic connections with various quality of a product. Research of model of electrothermal reactor installation with generation of low-temperature plasma is of special interest for synthesis of oxidic powder with the raised quality indicators. This approach defines technology of receiving details from powder materials of modern production.

1. Introduction

Synthesis of uniform oxidic powder in low-temperature plasma is of special interest for the production technology of difficult oxidic chemical compounds and drew attention of many researchers because of a wide range of use in various areas of the industry. Realization of technology of synthesis of powder in plasma substantially allows increasing quality of ferritic powders. This approach allows to solve many production problems. This decrease in multistaging of technological process, its duration, increases of indicators of quality, etc. [1].

The conducted researches in this direction are caused by [2] relevance of synthesis of oxidic powder in nonequilibrium low-temperature plasma with the set quality indicators. The presented results allow to judge the correct approach in the solution of an objective. This way based on application of a plasmatron with the nonequilibrium gas category with liquid electrodes was investigated by a number of authors [1, 3], however in their results properties of the received oxidic powders aren't presented.

Therefore the purpose of researches is synthesis of oxidic powder in nonequilibrium low-temperature plasma with the increased quality.

2. Experiment delivery.

On the basis of the offered model of electrothermal installation for powder synthesis, with the developed mathematical model of process of interaction of plasma with a liquid electrode installation was developed.

The electric discharge between firm and liquid electrodes is a source of low-temperature plasma. It allowed to receive optimum values of parameters of oxidic powder [1-5].

In figure 1 electrothermal installation for powder synthesis is presented. In the photo 1, and the electric discharge between the metal anode (materials of electrodes were used various became) and the liquid cathode is visible (various chemical compositions of electrolyte were used). In the photo 1, the stable level of electrolyte on z coordinate is shown.

Distinctive feature of the digit camera is its spherical form with the device of stabilization of level of electrolyte. This approach allows to receive stability of parameters of an electric discharge that defines quality of oxidic powder.

Researches were conducted with an atmospheric pressure. Interelectrode distance within 0,01 - 0,03 m [6, 7].

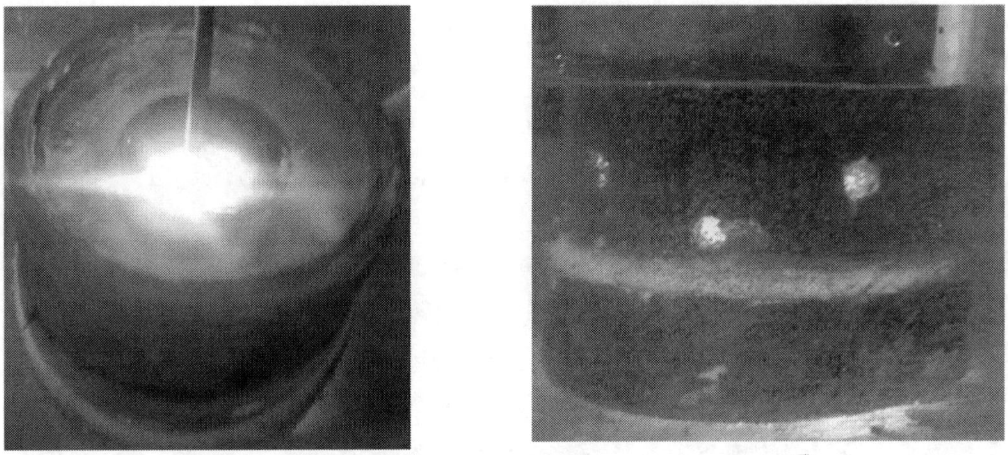

a) б)

Figure 1 - The photo of electrothermal installation with generation of low-temperature plasma for synthesis of oxidic powder

3. The results of experimental studies.

The analysis of the received oxidic powders showed dependence of their geometrical sizes on category current density. In figure 2 plasma with optimum value of density of current for obtaining the demanded quality of powder is visually presented.

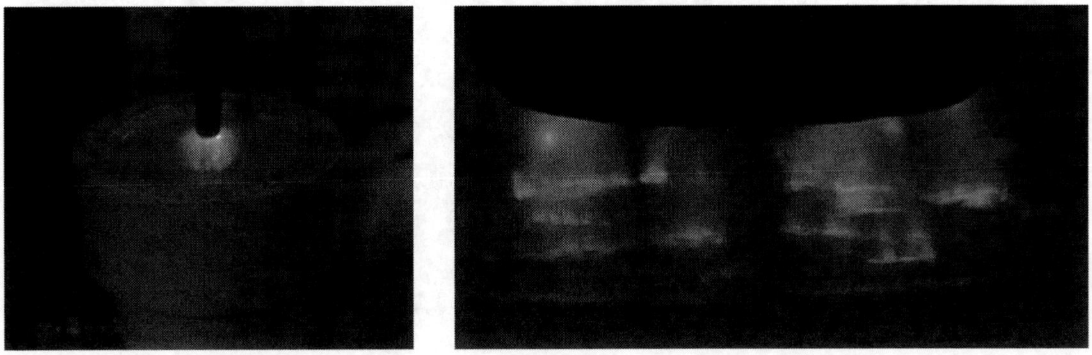

a) б)

Figure 2 - A shaping of low-temperature plasma between a firm and liquid electrode at various increase

In the photo 2, and the formed plasma at values of current is visible 400 мA. The enlarged photo (fig. 2, b) shows a mnogokanalnost of the received electric discharge. The number of channels is defined by the speed of rotation of the anode of a plasmatron. And it depends on the size of current of an anchor of the engine of a direct current.

The analysis showed that increase of density of current results in instability of indicators of quality of an output product. Process of synthesis of powder stopped with a diameter of shaping more than 10-2 m because of a big manufacturing defect. Formation of large particles it is caused by an electrode overheat under the influence of plasma at the big density of current.

Formation of powder begins with emergence of the gas category between a firm electrode and electrolyte. With increase in density of current on an electrode the product yield of technological process increases in the beginning, and then reaches saturation (for steel at iT = 105 And/m). Further increase of iT leads to fusion lower a crescent crown of an electrode, i.e. temperature in a contact spot on an electrode becomes higher than melting temperature Tpl, and the melted oxide of metal comes off an electrode, getting a sphere form at the expense of force of a superficial tension of a drop with a diameter from 10-6 to 5 - 10-4 m (fig. 3). The size of fractions of oxidic powder makes from 100 microns to 500 microns.

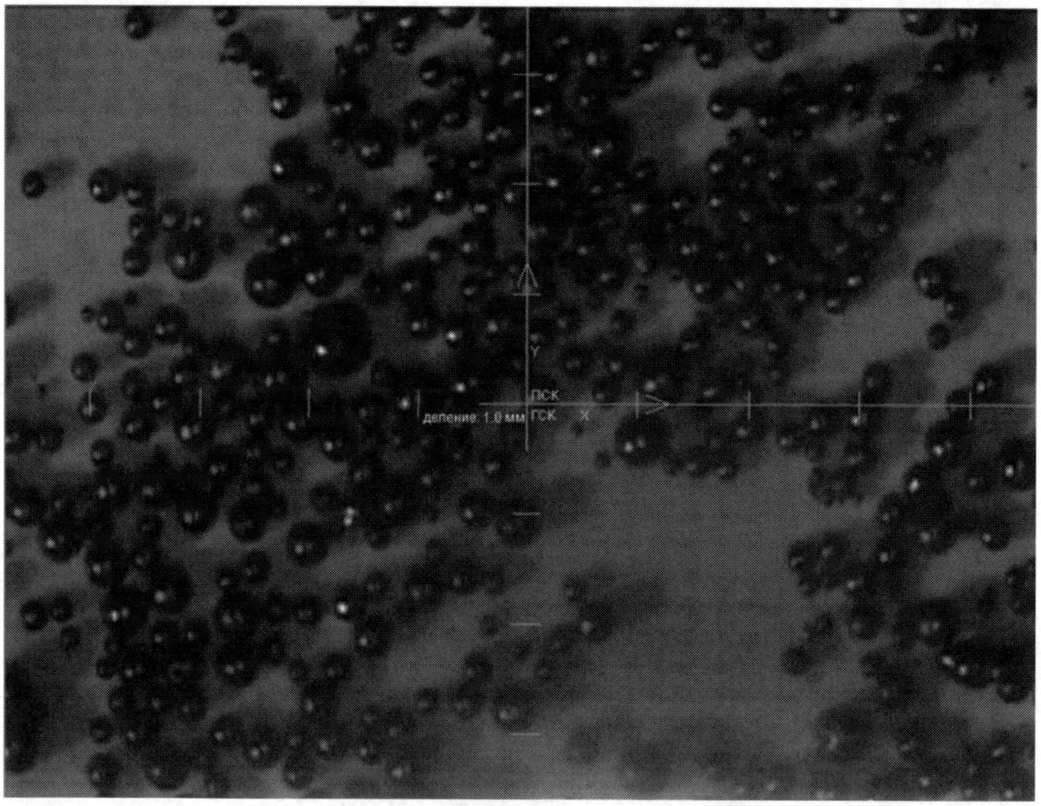

Figure 3 – The photo of the received oxidic powder

In figure 4 the photo of an end face of an electrode is shown. The erosion of a surface defines an exit of the final product, and the condition of a surface depends on a chemical composition of metal of an electrode.

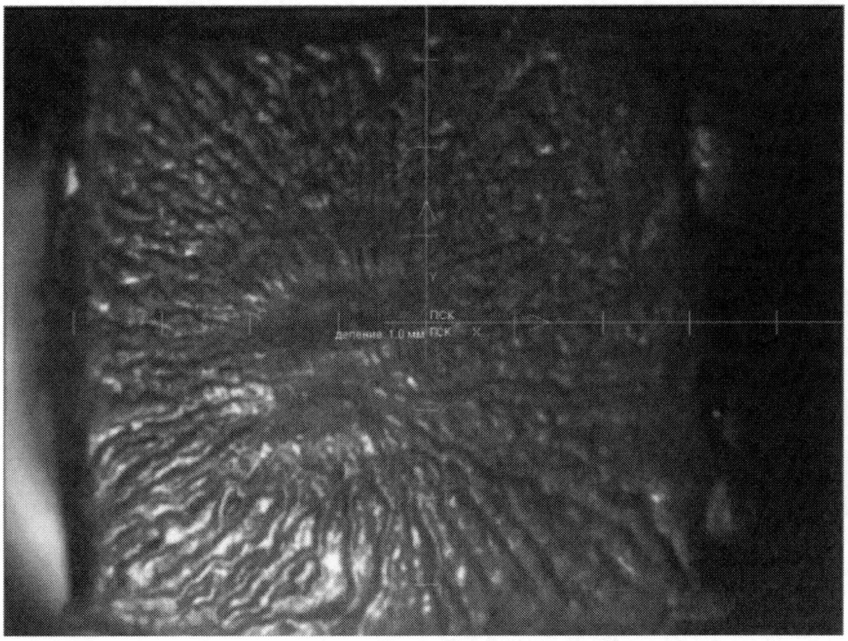

Figure 4 – Photo electrode tip

4. Conclusion.

The way of synthesis of oxidic powder in nonequilibrium low-temperature plasma showed increase of indicators of quality of process. It is of special interest for technology of the difficult oxidic chemical compounds possessing special electrophysical and physicomechanical characteristics. Realization of this way substantially allows to increase quality of a product for receiving ferritic powders.

References

[1] Gaysin, F.M. Steam-air categories between the electrolytic cathode and the metal anode with an atmospheric pressure Electric discharges in the steam-gas environment with nonconventional electrodes / F.M.Gaysin, E.E. Son//the Encyclopedia of low-temperature plasma / Under the editorship of Fortov V. E. M.: Science, 2000. – 241 pages.

[2] Peter Bruggtman and Cristophe Leys. Not thermal plasmas in liquids and kontragiruyushchy with liquids (the thematic review) / P. Bruggtman, Cr. Leys//J. Phys. D: Appl. Phys. 42 (2009) 053001 (28 pp).

[3] P. Mezei and T. Ceserfalvi. Electrolyte cathode atmospherie glow dischartsyges for direct solution analysis Appl. Spectrosc. Rev42 (2007), S. 573 – 604.

[4] Gaysin F.M., Dream E.E., Shakirov Yu.I. The volume category in the steam-gas environment between firm and liquid electrodes / - M.: Publishing house of VZPI, 1990. - 90.

[5] Bagautdinova L.N., Gaysin F.M. The multichannel category in the carrying-out liquid with an atmospheric pressure thermophysics of high temperatures. 2010. - 135.

[6] Haybullin I.I., Saubanov R. R., Israfilov I.Kh., Rakhimov R. R. Perspective use of the high-concentrated energy for superficial heat treatment a product / the On-line electronic scientific and technical magazine "Socially — Economic and Technical Systems" No. 1(58) Naberezhnye Chelny: INEKA, 2011. http://sets.ru.

[7] Saubanov R. R., Haybullin I.I., Gaysin F.M. Synthesis of oxidic powder in nonequilibrium low-temperature plasma with nonconventional electrodes / the 21st Century. Fundamental science and technologies – Moscow, 2013. – 189-191.

Gas discharge with liquid electrolyte cathode in the mode of occurrence of the constricted channels

Kh K Tazmeev and A Kh Tazmeev

Branch of Kazan Federal University in Naberezhnye Chelny, 68/19 Mira Street, Naberezhnye Chelny, 423810, Russia

E-mail: tazmeevh@mail.ru

Abstract. Gas discharge between liquid electrolyte and a copper electrode at capacities that make up tens of kilowatts is experimentally investigated. As the liquid electrolyte we used the aqueous solution of sodium chloride. Identification of conditions of occurrence of the constricted channels. The electrical and spectral characteristics of the discharge in such conditions were studied.

1. Introduction

When using a liquid electrolyte as the cathode plasma is formed mainly of liquid-phase substance. Plasma properties and characteristics of the gas discharge is directly dependent on the chemical composition of the electrolyte, which is confirmed by a number of experimental studies conducted with various electrolytes in a wide range of current, interelectrode spacing, and other parameters [1-6]. By varying the electrolyte composition can be obtained plasma containing different chemically active components. The aim of this work was to study the characteristics of the gas discharge with liquid electrolyte cathode at occurrence a constricted channels that serve as additional sources of chemically active species, such as excited atoms, ions and electrons.

2. Experiment

The experimental setup is shown in figure 1. As the electrolyte were used saline solutions in distilled water. The electrolyte is flowed from the tank *1* in the form of a cylinder of diameter D = 75 mm in the vertical upward direction (figure 1). The current-carrying element was a graphite plate *2* placed inside the cylinder. Discharge was burning between the electrolyte *3* and the copper anode *4*, cooled with tap water. The length *l* of the discharge gap was set in the range of 30 to 60 mm. On graphite plate and an anode was applied voltage from three-phase full-wave rectifier with an output voltage 3200 V. The pulsations voltage was smoothed with using LC-filter. Current regulation the discharge was carried by a stepwise change of the ballast resistor in the circuit of the electric power.

The concentration *C* and the electrical conductivity σ of the electrolyte was measured with conductivity meter ANION 4150. The current *I* and the voltage *U*, applied to the plate *2* and the anode *4*, were recorded with the oscilloscope AKIP-15/1. Instant images of the discharge were photographed with high-speed camera VIDEOSCAN-415.

The emission spectra were recorded using high-speed fiber-optic spectrometer AvaSpec-3648 in the wavelength range 484-708 nm with a resolution of 0.15 nm (diffraction grating 1200 lines / mm, the input optical gap of 10 microns). The image of the plasma column using converging lens *L* projected entirely on the input fiber optic cable. High-speed video camera *5* and spectrometer *6* were operable associated with computers *7* and *8*.

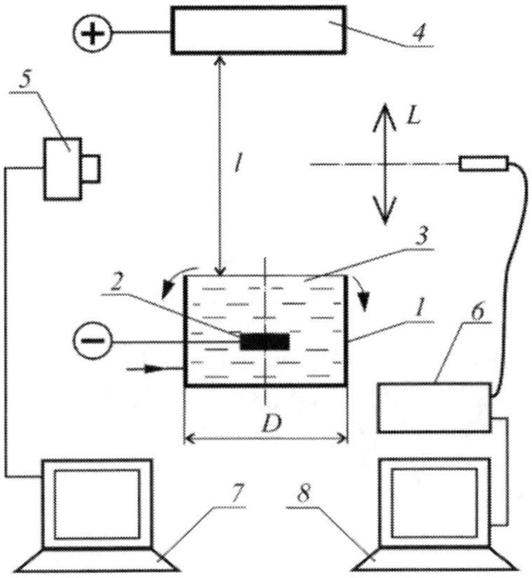

Figure 1. The experimental facility.

3. Results and their discussion

Emergence constricted channels in the discharge region contribute to reducing the concentration (respectively, conductivity) of the electrolyte. Figure 2 shows photographs taken at the time of occurrence of such channels.

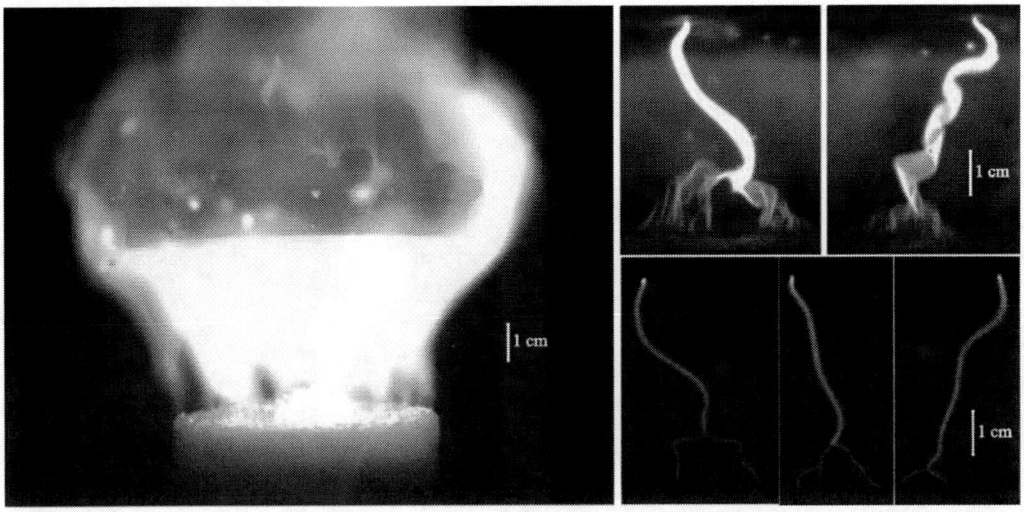

Figure 2. General view of a gas discharge (left photo) and constricted channels: $I = 12 \pm 0{,}5$ A, $l = 5$ cm, $\sigma = 2{,}0 \pm 0{,}1$ mS/cm ($C = 1{,}0 \pm 0{,}1$ g/l), the exposure of 196 μs (the top row of the right) and 3 μs (bottom row of the right).

On the general form of a gas discharge, which was photographed with the maximum aperture, constricted channel stands out as a bright area (figure 2, left photo). Shots taken with the closed diaphragm, have revealed the range of light images. As can be seen, unlike a yellow background radiation generated by sodium, constricted channels are allocated in a light-colored red (figure 2, the top row of the right). The image of the constricted channel is obtained in the form of complex, sinuous shapes with branches resting on the surface of the electrolyte. In the footage with less exposure, observed the same color throughout the length of the constricted channel. From the monotony of color implies that in the channel are numerically dominated by the excited particles of one kind. The most likely candidates for the role of such particles are hydrogen atoms because hydrogen gas is released at the cathode electrolysis processes and flows into the discharge region.

The photographs shown on the middle and lower ranks of figure 2, made with different exposure more than 60 times. Comparing them, we can conclude that the complex intricate images constricted channels (frames with a larger exposure) were formed with of stacking simple shapes (frames with less exposure).

The presence of radiating of hydrogen atoms have confirmed by spectral studies. In the emission spectrum of the gas discharge are the Balmer lines of hydrogen H_α and H_β, and at the moment of appearance of the constricted channels line H_α acquired much more intense yellow D-line of sodium (figure 3).

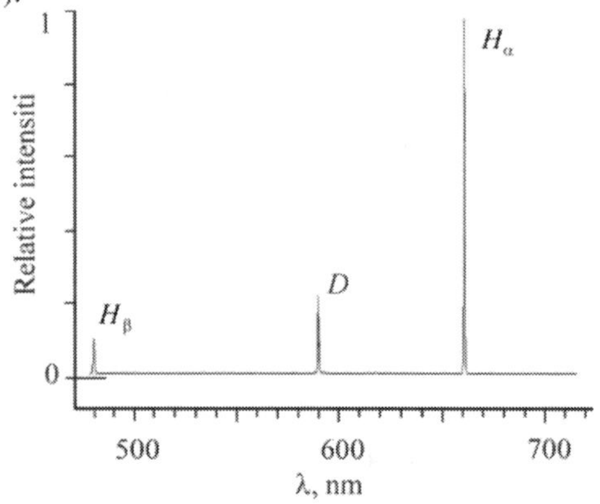

Figure 3.The emission spectrum of the gas discharge.

Since the appearance and disappearance constricted channels occurs almost instantaneously, spectra were recorded without averaging with a minimum integration time in maximum performance mode spectrometer. In particular, for the spectrum shown in the figure 3 as an example, the integration time was 0.25 ms.

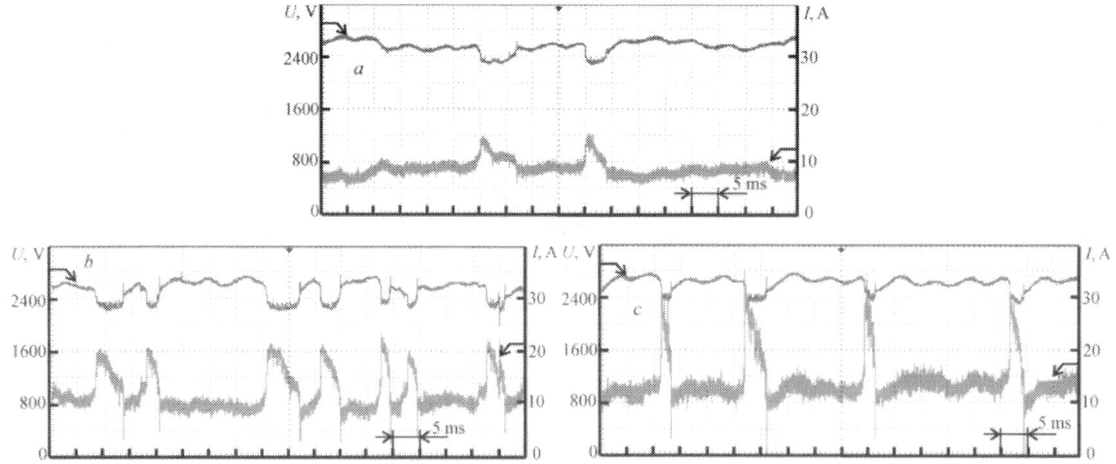

Figure 4. Oscilloscope traces of currents and voltages: l = 5 cm, σ = 2,4 ± 0,1 mS/cm (C = 1,2 ± 0,1 g/l), a – $R_{ballast}$ = 52 Ω, b – 34, c – 14,5.

As follows from the oscillograms, parameters constricted channels vary significantly with decreasing ballast resistance. They occur more frequently and the maximum current through a single channel increases. At the same time in the change $\Delta\tau$ any patterns not explicitly detected. In the moments of occurrence of the constricted channels voltage U, registered between the anode and the current-carrying graphite plate is reduced.

An increase in current occurs because in case of constricted channel decreases the electrical resistance of the discharge region, respectively, and the total resistance of the circuit. For the same reason, i.e. because of the reduction in the total electrical resistance increases the maximum current in the constricted channel while reducing ballast $R_{ballast}$. The presence of the emission spectrum of the hydrogen Balmer lines allows us to estimate the electron temperature T_e of the plasma in the constricted channel. As an example, the table shows the values of T_e, calculated by the method of relative intensities. The emission spectrum of gas discharge used under the conditions: l = 5 cm, σ = 2,4 ± 0,1 mS/cm (C = 1,2 ± 0,1 g/l).

Table. The electron temperature T_e.

τ_{int}, ms	T_e, K				
	1	2	3	4	5
0,12	4926	4980	5039	5178	5468
0,25	4209	4500	5091	5367	5677
0,50	4993	5039	5340	5747	6282
1,00	4298	5059	5607	5835	6451

Calculated values T_e fluctuate in considerably wide range. Apparently, the reason for such a wide difference in the T_e is that the signal integration spectrometer does not always coincide in time with the time of point of maximum discharge current. According to the calculated data the maximum current in the constricted channel is within the 5500 - 6500 K.

4. Conclusions

The experiments showed that the gas discharge burns in the mode of occurrence of the constricted channels continuously without time limitations. The power released in the discharge is in the range of tens of kilowatts. Thus, from the experiments it follows that the gas discharge in the combustion mode with constricted channels may serve as a plasma source for energy-intensive technologies.

References

[1] Sluginov N P 1878 *Jurnal Russkogo phiziko-himicheskogo obchestva V.* **10,** *vipusk 8* 241

[2] Valiev R A, Gaisin F M, Romanov E S and Shakirov Yu I 1991 *Physics and chemistry of materials treatment* **25** 654

[3] Webb M R, Andrade F J , Gamez G, McCrindle R and Hieftje G M 2005 *J. Anal. At. Spektrom.* **20** 1218

[4] Maksimov A I and Khlyustova A V 2009 *High Energy Chemistry* **43** 149

[5] Bruggeman P and Leys C 2009 *J. Phys. D: Appl. Phys.* **5** 053001

[6] Bobkova E S, Grinevitch V I, Isakina A A and Rybkin V V 2011 Izvestia visshih uchebnih zavedenii. Seria.: Himia i Himicheskih tehnologii **6** 3

Longitudinal distribution of electrical parameters in normal glow discharge

B A Timerkaev, M M Ahmetov, B R Zalyaliev, O A Petrova and D I Israfilov

KAZAN NATIONAL RESEARCH TECHNIKAL UNIVERSITY named after A.N.TUPOLEV - KAI, 10, K.Marx Str., Kazan, Republic of Tatarstan 420111 RUSSIA

e-mail: of@kai.ru

Abstract. The behavior of the glow discharge at low pressure is considered.

1. Introduction

In this paper we propose a combined experimental-theoretical method for determining the concentration distributions of electrons and ions in the normal glow discharge. The distribution of potential φ on length of the discharge chamber is measured experimentally.

2. Exterimantal

The discharge chamber for experimental studies consisted of a water-cooled copper electrodes, interelectrode insert made of quartz glass and a device for probe measurements. The potential was measured using a single probe consisting of a tungsten wire with a diameter of 0.05 mm, the insulating quartz capillary and the quartz tube. The potential distribution was measured in the evacuated discharge chamber with a glow discharge in the air. The discharge mode with cathode spot completely filling the cathode was used for stabilization of the discharge area and the discharge stayed symmetric despite the presence of the probe. The current value of this mode was 20 mA. The probe position along the discharge chamber from the cathode to the anode could be changed by a micrometer screw with a pitch of 1mm.

3. Results

The results of experiments of measurement of the potential distribution along the axis of the discharge in the air at a pressure of 3 Torr are shown in Figure 1. The discharge voltage was 520 V. Most part of the voltage falls at the cathode layer of thickness of 0.2 cm. After the cathode fall the potential profile graph has slow potential variation area up to 0.45 cm at $p = 3$ Torr. From this place the constant field is restored until the anode layer.

The principle in this case is the presence of weak electric field potential variation area that is corresponding to the minimum of the electric field strength.

The experimental data was approximated by the least squares method. The whole area was divided into several sections for the best fit with the experimental data.

Under these conditions approximating function $\varphi(x)$ has the following form:

Journal of Physics: Conference Series **567** (2014) 012036 doi:10.1088/1742-6596/567/1/012036

$\varphi(x) = 254{,}626[1 - exp(-31{,}2816\,x)]$, $0{,}07\text{cm} > x \geq 0$

$\varphi(x) = (1.48466 \cdot 10^{11} (8.12255 \cdot 10^{-7} x^2 - 0.000021125\,x^3 + 0.000286649\,x^4 - 0.00250675\,x^5 + 0.015211\,x^6 - 0.0662176\,x^7 + 0.209786\,x^8 - 0.484373\,x^9 + 0.806179\,x^{10} - 0.941637\,x^{11} + 0.732109\,x^{12} - 0.340138\,x^{13} + 0.0714286\,x^{14})) + 115$,
$0{,}538\ \text{cm} \geq x \geq 0.070\ \text{cm}$,

$\varphi(x) = 223.33674 + 88.40096\,x$, $2{,}663\ \text{cm} \geq x > 0{,}538\ \text{cm}$,

$\varphi(x) = (2.82108 \cdot 10^9 x^2 - 5.60553 \cdot 10^9 x^3 + 5.47704 \cdot 109\,x^4 - 3.25889 \cdot 10^9 x^5 + 1.26128 \cdot 10^9 x^6 - 3.21059 \cdot 10^8 x^7 + 5.20968 \cdot 10^7\ x^8 - 4.90303 \cdot 10^6\ x^9 + 204230\,x^{10}) - 4.50364 \cdot 10^8)$,
$3\ \text{cm} > x > 2{,}663\ \text{cm}$.

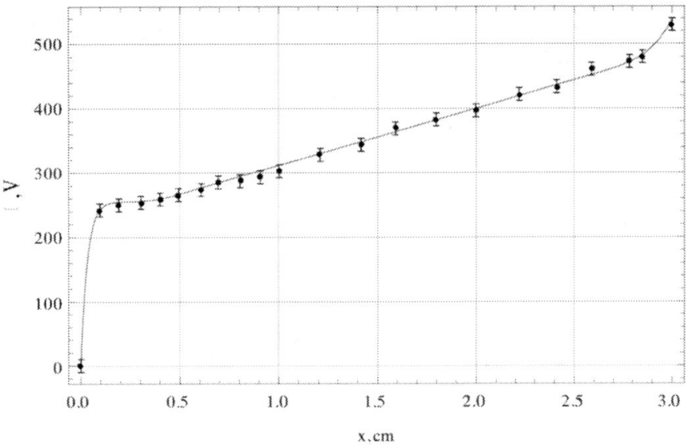

Figure 1. The longitudinal distribution of the potential along the discharge chamber, $p = 3$ Torr.

Field strength was calculated as the derivative of the function. The function $E(x)$ obtained this way was used in subsequent calculations of electrons and ions concentrations.
The graph of strength E dependence on the distance x from the cathode is shown in Figure 2. The curve has conspicuous minimum near $x = 0.2$ cm. Such a strong dependence $E(x)$ in this area can not be obtained solving the system of equations (1) - (5).
The theoretical model of the glow discharge consist of a standard set of equations, based on the Poisson equation and the continuity equation for the current density of electrons j_e and ions j_p considering the volume recombination

$$\frac{dj_e}{dx} = \alpha j_e - \beta e n_e n_p \ , \quad (1)$$

$$\frac{dj_p}{dx} = \alpha j_e - \beta e n_e n_p \ , \quad (2)$$

$$= -\frac{dE}{dx} = \frac{e}{\varepsilon_0}(n_e - n_p)\ , \quad (3)$$

$$j_p = e n_p \mu_p E, \quad (4)$$

$$j_e = e n_e \mu_e E\,, \tag{5}$$

where $\alpha = Ap\exp(-\frac{Bp}{E^*})$ - the Townsend coefficient, μ_e and μ_p - mobility electrons and ions, respectively, β - volume recombination coefficient, φ - electric field potential, p - pressure. A, B-electrophysical parameters, for nitrogen they have values [1] A = 12 (cm·Torr)$^{-1}$, B = 342 V / (cm·Torr)$^{-1}$; e = 1,6· 10^{-19} C - electron charge, ε_0 = 8,85· 10^{-12} F/m - electric constant, E - the electric field, E * - non-local value of the electric field determining the rate of ionization, $\gamma = \frac{j_e(0)}{j_p(0)}$ - the coefficient of secondary ion-electron emission, determining the ratio of the electron j_e (0) and ion j_p (0) currents at the cathode.

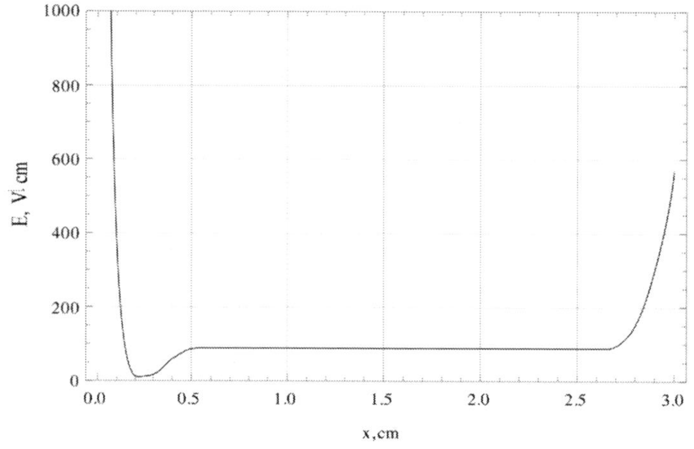

Figure 2. The dependence of the electric field strength on the distance from the cathode.

Nonlocality of the dependence of ionization rate on the electric field E was taken into account by the method described in [2]:

$$E = E - \frac{dE}{dx}\frac{\lambda_{ion}}{2}\,, \tag{6}$$

λ_{ion} - electron mean free path between collisions:

$$\lambda_{ion} = \frac{1}{\sqrt{2}\sigma_i N}\,, \tag{7}$$

σ_i - single ionization cross-section, $N = \frac{p}{kT}$ - concentration of neutral atoms at a temperature T, k = 1,38 10^{-23} J/K - Boltzmann constant. The total current density j_0 is constant at any cross section of the discharge tube:

$$j_0 = j_e + j_p\,. \tag{8}$$

To determine the concentration distributions of electrons and ions in the discharge chamber from the cathode to the anode is necessary to solve the system of equations (1) - (5) with the appropriate boundary conditions.

This system of equations is rigid and its numerical solution requires the use of special methods. The use in this paper of experimental dependences of the potential distribution *φ(x)* and intensity *E(x),* let avoid the problems associated with the rigidity of the system of equations.

The set of equations (1) - (5) can be reduced to a differential equation for the electron density n_e

$$\frac{dn_e}{dx}E + n_e\frac{dE}{dx} = Apn_e E\exp(-\frac{Bp}{E}) - \frac{\beta}{\mu_e}n_e n_p \quad . \tag{9}$$

The ion concentration was calculated using expressions for current densities (4), (5) and condition of constancy of the total current density (8):

$$n_p = \frac{j_0}{e\mu_p E} - n_e\frac{\mu_e}{\mu_p} \quad . \tag{10}$$

Equation (9) was solved numerically. The derivative of the experimental data of the potential distribution *φ(x)* was use as strength *E(x)*. The concentration of the ions was calculated by equation (7) with use of values of the electron concentration. It was taken into account that the n_e (0) = 10^7 cm^{-3}, j_0 = 0,013778 A/cm^2. The mean free path of electrons was calculated by the formula (9) with $p = 3$ Torr, λ_{ion} = 0,0103 cm.

The numerical values of the other parameters used in the calculation are [1]: A=12(cm·Torr)$^{-1}$, B = 342 B/(cm·Torr)$^{-1}$, μ_e p = 4,4·10^5 Torr·cm^2/(V·s), μ_p p = 1,45·10^3 Torr·cm^2/(V·s), β = 3,2·10^{-7} cm^3/s, γ = 0,1.

The obtained dependences of the electron and ion concentrations n_e and n_p for pressure 3 Torr are shown in Figure 3.

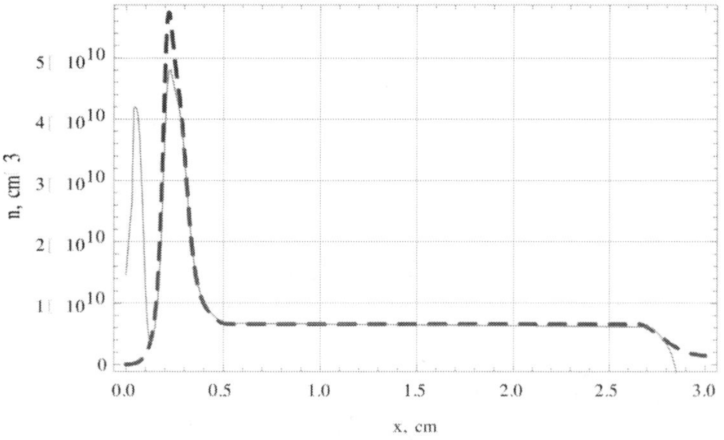

Figure 3. Calculated dependences of the electron (dashed line) and ion densities (solid line) for the pressure 3 Torr, $n_e(0)$ = 10^7 cm^{-3}, j_0 = 0,013778 A/cm^2.

As shown in Figure 3, the discharge has its self-organization after its ignition: distributions of the electric field potential and strength, as well as the concentrations of electrons and ions gain the appropriate form. The concentration of electrons has sharp rise practically at the electron mean free path. Physically, this is due not only to the processes of ionization due to electron impact, but to their drift velocity slowdown, since in this region the electric field intensity decreases sharply. Strange behavior of the ion concentration near the cathode is

also a due to the discharge self organization. The first graph decreasing to the cathode side of the main peak is due to the increase in the drift velocity of the ions. In this area, growth of their concentration due to ionization of neutral atoms by electron impact is still negligible.
But the further emergence of ions due to ionization begins to prevail over their drift leaving. Ion concentration begins to increase. But here comes a new twist - the electrons drift leaving again becomes prevalent. The resulting diagram of the distributions of concentrations of charged particles completely describes the layered structure of the glow discharge. Region from the cathode to the first peak of concentration of positive ions is Aston dark space. Next area to the region of the first minimum of concentration of positive ions is cathode glow. Next area, including peak concentration of charged particles, is Faraday dark space. From the above analysis it is clear that the growth of the concentration of charged particles in the Faraday dark space is not due to intense ionization by fast electrons from the cathode region, but due to their drift motion slowdown.

5. Conclusions

Thus, the use of the experimental curve of the potential distribution along the discharge chamber and considering the nonlocal dependence of the Townsend coefficient on the electric field strength yielded distribution of the electric field and to determine the exact nature of changes in concentration of charged particles along the axis of the discharge. Nonlocal character of the Townsend coefficient on the electric field causes the experimentally observed "downfold plot" (plot with a small field change) in the potential distribution and defines a layered structure of the glow discharge.

References

[1] Kudryavtsev A A , Smirnov A S and Tsendin L D 2010 *Fizika tlejushego razrjada* Spb M Krasnodar Lan.

[2] A I Saifutdinov and B A Timerkaev 2012 *Journal of Engineering Physics and Thermophysics* **85** 1202.

Nonlinear oscillations of gas in an open tube near the resonance frequency in the shock-free mode

L A Tkachenko[1,2] and M V Sergienko[2]

[1] Institute of Mechanics and Engineering, Kazan Science Center, Russian Academy of Sciences, 2/31 Lobachevsky str., Kazan 420111, Russia
[2] Kazan (Volga Region) Federal University, 18 Kremlyovskaya str., Kazan 420111, Russia

E-mail: luda_tkachenko@inbox.ru, maria_sergienko@mail.ru

Abstract. The forced oscillations of gas in an open tube, excited by harmonical oscillations of piston in the shock-free mode were investigated near the first first eigenfrequencies. An expression for the pressure oscillations of gas was obtained for the tube with unrounded end without flange. The amplitude impact of piston displacement on the oscillations of pressure and velocity of the secondary flow of gas was investigated. The comparison of theoretical calculations with experimental data was executed. The effect of secondary flow on the particle drift along the tube axis with acoustic oscillations of gas was shown.

Introduction

Various modern equipment often contain tubing systems with finite length with rest or moving homogeneous or heterogeneous operating environment [1]. Review of studies on nonlinear resonant oscillations of a homogeneous gas in tubes with different conditions at the ends is given in [2]. Focused on the study of nonlinear oscillations of gas in the tube and the external wave field at large excitation amplitudes (of the order of 0.1-0.4 bar) in the shock-wave mode, periodic shock waves was occurred. Gas oscillations in the shock-free mode at low amplitudes of excitation (about 0.01 bar) is not well understood.

The purpose of this work is research of the resonant oscillations of gas occurring by secondary flow in the shock-free mode in a tube with an open end.

1. Basic conditions

Harmonic oscillations of gas $x = l\sin\omega t$ in a long cylindrical tube ($l << L_0, L_0/R >> 1$, where l - the amplitude of the displacement of the piston, R - radius, L_0 - length, ω - the angular frequency) are described by [3]

$$\overline{p}_1 = r_1\cos\left(k_1x + \alpha_1 + i\beta_1\right)e^{i(\omega t+\psi_1)},\ \ \overline{u}_1 = -ir_1\sin\left(k_1x + \alpha_1 + i\beta_1\right)e^{i(\omega t+\psi_1)}, \qquad (1)$$

[1] To whom any correspondence should be addressed.

Where $\bar{p}_1 = p_1/\rho_0 c_0^2$, $\bar{u}_1 = u_1/c_0$, $k_1 = (\omega/c_0)(1+\beta_1' - i\beta_1')$ – complex wave number, $\beta_1' = \sqrt{\mu/2\rho_0\omega}\left[1+(\kappa-1)/\sqrt{\Pr}\right]/R$ – the dissipation coefficient, $\kappa = c_p/c_v$ – ratio of specific heats at constant pressure and volume, ρ_0 – density, Pr – Prandtl number, c_0 – speed of sound in the undisturbed gas, r_1, ψ_1, α_1, β_1 – constants.

Substituting (1) into the boundary conditions at the ends of the tube [4]

$$\bar{u}_{1p} = \mathrm{M}_p e^{i\omega t},\ x=0;\ \bar{p}_{1E} = m r_1^2 e^{i(\omega t+\varphi_1)},\ m = \frac{2}{3\pi}(1+k_{in}),\ x=L. \quad (2)$$

Here, m contains parameter k_{in}, which is determined experimentally [5], and depends on the shape of the open end of the tube, $L = L_0 + \sigma R$ - reduced tube length, σR - correction of Rayleigh, σ =0.6133. As a result, we obtain expressions for the constants r_1, ψ_1, α_1, β_1

$$r_1\sqrt{\cos^2\left[k_0 L(1+\beta_1')\right] + (k_0 L\beta_1' + m r_1)^2 \sin^2\left[k_0 L(1+\beta_1')\right]} = \mathrm{M}_p,$$

$$\alpha_1 = \frac{\pi}{2} - k_0 L(1+\beta_1'),\ \beta_1 = k_0 L\beta_1' + m r_1,\ \psi_1 = \mathrm{arctg}\left(\frac{\mathrm{tg}\,\alpha_1}{\beta_1}\right), \quad (3)$$

2. Experiment

Experimental researches were carried out on an automated equipment, previously used to study the nonlinear oscillations of a homogeneous gas in a closed tube in the shock-free mode [6]. Longitudinal stationary oscillations of gas in a vertical quartz tube (length L_0 = 1.06 m, inner diameter $2R$ = 0.0365 m), created by a sinusoidally oscillating piston. Piston is driven by the vibration generator with power amplifier TV51075. Passive end of the tube was opened. Vibration generator was controlled by piezoelectric IEPE accelerometer controller using a software module SineVIEW (VR610), installed on a computer. Piezoelectric pressure sensor was hostead close to the piston, the signal from sensor was transfered through the three-channel bridge voltage amplifier to the digital oscilloscope and then to the computer, where the value and form of signal were recorded with accurate to 0.3 mV.

3. Results

The theoretical dependence and the experimental data for the dimensionless amplitude of oscillations of the gas pressure are given on the Figure 1, $\Delta p_1 = (\Delta p_1/p_0)\times 10^2$ (Δp_1 – difference of the maximum and minimum values of the pressure for the piston stroke, p_0 – the equilibrium pressure of of gas) on the relative frequency of excitation $\bar{\nu} = \nu/\nu_1^?$ ($\nu_1^?$ –experimentally observed resonance frequency) for a given values of the relative amplitude of the displacement of the piston. There is dotted line represents the resonant frequency of excitation of the gas, calculated from the linear theory [2] $\nu_1 = c_0/2L$=80.1 Hz. Dashed line is a frequency obtained in the presence of losses $\nu_1^* = c_0/2L(1+\beta_1')$=78.6 Hz, which mathes with the value observed in the experiment. The results of theoretical oscillation calculations of gas pressure for the unrounded end of tube without flange m = 0.84 [5] are somewhat higher than the experimental data. The discrepancy is about 10% for all studied displacement amplitude of the piston. Also Figure 1, b shows a typical waveform recordings with pressure oscillations of gas in the time of the experimentally observed resonant frequency. Pressure diagrams have an uninterrupted view, and no breaks were observed. The form of diagrams is somewhat different from the harmonic (the duration of the leading edge somewhat less than the duration of the trailing edge of the wave).

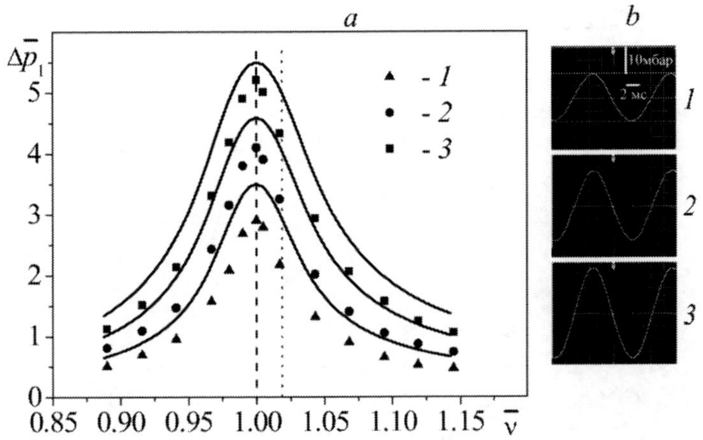

Figure 1. Dependences of gas pressure near the piston on $\bar{\nu}$ (a) and on t (b): $1-\bar{l}=6.13$, $2-\bar{l}=9.43$, $3-\bar{l}=12.73$. Solid lines - theory, dotted - experiment.

To calculate the secondary flows we'll use the method proposed in [7]. As a result of integration, we have the expressions for the axial and radial components of the velocity of the secondary current in the form

$$\langle u_2\rangle=\frac{c_0r_1^2}{4}\left\{\left[g_1(0,\mathrm{Pr})\left(1-\frac{2r^2}{R^2}\right)+g_1(\eta,\mathrm{Pr})-\frac{\delta}{R}g_3(0,\mathrm{Pr})\left(1-\frac{r^2}{R^2}\right)\right]\sin 2v+\right.$$
$$\left.+\left[g_0(0,\mathrm{Pr})\left(1-\frac{2r^2}{R^2}\right)+g_0(\eta,\mathrm{Pr})+\frac{\delta}{R}g_2(0,\mathrm{Pr})\frac{1}{\beta_1'}\left(1-\frac{r^2}{R^2}\right)\right]\mathrm{sh}\,2w\right\}, \tag{4}$$

$$\langle \upsilon_2\rangle=\frac{\omega R r_1^2}{8}\left\{\left[2g_1(0,\mathrm{Pr})\left(\frac{r^3}{R^3}-\frac{r}{R}\right)-\frac{\delta}{R}g_3(0,\mathrm{Pr})\left(\frac{r^3}{R^3}-\frac{2r}{R}\right)-\frac{\delta}{R}g_3(\eta,\mathrm{Pr})\right]\cos 2v-\right.$$
$$\left.-\left[2\beta_1'g_0(0,\mathrm{Pr})\left(\frac{r^3}{R^3}-\frac{r}{R}\right)+\frac{\delta}{R}g_2(0,\mathrm{Pr})\left(\frac{r^3}{R^3}-\frac{2r}{R}\right)-\frac{\delta}{R}g_2(\eta,\mathrm{Pr})\right]\mathrm{ch}\,2w\right\}, \tag{5}$$

where $v=\frac{\omega x}{c_0}(1+\beta_1')+\alpha_1$, $w=\beta_1-\frac{\omega x}{c_0}\beta_1'$ and $g_0(\eta,\mathrm{Pr})$, $g_1(\eta,\mathrm{Pr})$, $g_2(\eta,\mathrm{Pr})$, $g_3(\eta,\mathrm{Pr})$ are the functions depending on $\eta=(R-r)/\delta$, κ and Pr, $\delta=\sqrt{2\mu/\rho_0\omega}$ - is the thickness of the Stokes's layer.

The calculations show that the flow in the form of one toroidal vortex with movement direction to the open end of the tube is being observed at the tube axis. In the near-wall region there are vortices with movement direction opposite to the one in the central region. One can note the presence of areas of the vortex movement near the piston and at the open end of the tube with the direction of movement opposite to the main vortices. The flow velocity in the near-wall region is essentially lower than in the stream nucleus (for the axial component of velocity the difference is 10 times, for the radial one – 100 times). This is connected with the presence of the near-wall absorption. While amplitude of the

Journal of Physics: Conference Series **567** (2014) 012037 doi:10.1088/1742-6596/567/1/012037

displacement of the piston increases both axial and radial velocity components increase as well. Figure 2 shows the distribution of the axial velocity component on the axis (with $r \to 0$) along the tube length for different amplitudes of the displacement of the piston. Near the piston and at the open end of the tube the value of the axial component of the velocity of the secondary flow differs from zero and has a negative value. This is connected with the presence of the item with sh $2w$ which is aperiodic on x in the expression for the axial velocity component. With the increase of the amplitude of the displacement of the piston the secondary flow velocity increases on the axis of the tube.

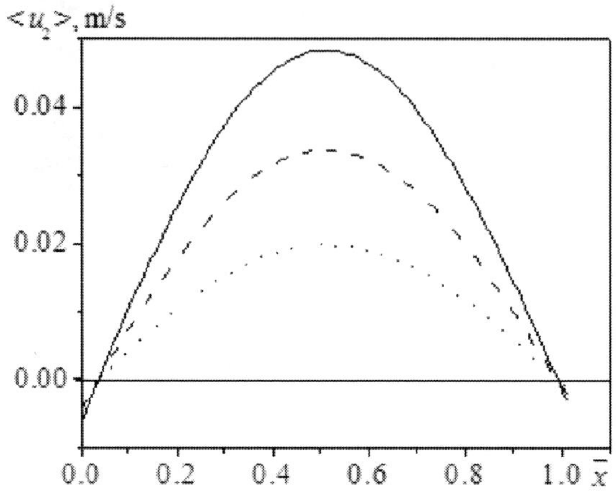

Figure 2. Dependences of $\langle u_2 \rangle$ on $\bar{x}$ with $\bar{r} = 0$: 1 – $\bar{l}$ =6.13, 2 – $\bar{l}$ = 9.43, 3 – $\bar{l}$ = 12.73

The presented calculation of the secondary flow velocity can be used to estimate the influence of the acoustic flow on the particle dynamics with the gas oscillations in tube. The Stokes' force, the Archimedes' force, the force of friction, the Basset's force and others, as well as acoustic flows [8, 9] act on a particle in a wave field. At the same time the direction and the drift velocity of the particle depend on the particles parameters and the wave field characteristics [10, 11], which is widely used in the wave technologies of the particles separation.

Consider the data for the experimental investigation of the drift of polystyrene flat particles (diameter 9 mm, thickness 0.4 mm, weight 5.7 mg) strung along the axis of the tube open at gas oscillations on the fishing line. Particle was placed at predetermined distances along the tube axis and after launching the sinusoidal oscillations particle drift started filming. Particle placed near the open end of the tube, moving up to a certain point in the direction of the piston, making oscillations. When placing the particles closer to the piston, reversed movement is observed to the open end of the tube. This behavior of particle is consistent with obtained pattern of secondary flows. Thus, when the relative amplitude is $\bar{l}$ = 16.34, the velocity of particle installed near the open end is 0.025 m / s, and installed near the piston - 0.05 m / s. The particle velocity is 2.5 times less than the velocity of the secondary flow on the axis of the tube. This is due to the influence of the other forces drift such as friction forces of the fishing line.

4. Conclusions

The forced oscillations of gas in an open tube, excited by harmonical oscillations of piston in the shock-free mode were investigated near the first eigenfrequencies frequency. The observed resonance frequency has less value than calculated by the linear theory, but coincides with the frequency calculated by taking into account the nonlinear theory of near-wall losses. We found that the maximum axial velocity component is in the middle of the tube, and radial component maximum is in the ends of the tube. The flow velocity in the boundary region is considerably lower than in the core stream (the difference for an axial velocity component is 10 times, for the radial is 100 times). This is due to the presence of a wall surface absorption. With increase of the amplitude of the piston displacement and intensity oscillations, there is increasing of the axial and radial velocity components. It is shown that the acoustic flow has a significant effect on the particles drift. Particle moves to a certain point in the direction of the piston, making oscillations near the open end of the tube. Reverse movement is observed in the open end of the tube when particle is placing closer to the piston.

Acknowledgments
This work was funded by the subsidy of the Russian Government to support the Program of Competitive Growth of Kazan Federal University among World's Leading Academic Centers.

References
[1] Ganiev R F, Ukrainskiĭ L E 2008 *Nonlinear Wave Mechanics and Technology* (Moscow: Nauchno-izdatel'skiĭ tsentr "Regulyarnaya i khaoticheskaya dinamika)
[2] Ilgamov M A, Zaripov R G, Galiullin R G and Repin V B 1996 *Appl. Mech. Rev.* **49**(3) 137
[3] Galiullin R G, Galiullina E R and Permiakov E I 1996 *Acoustical Physics* **42**(6) 769
[4] Tkachenko L A and Galiullin R G 2010 *Russian Aeronautics* (1) 31
[5] Sturtevant B B and Keller J J 1978 *ZAMP* 29(3) 473
[6] Tkachenko L A and Zaripov R G 2013 *Russian Aeronautics* (1) 28
[7] Galiullin R G, Timokhina L A and Filipov S E 2001 *Acoustical Physics* **47**(5) 529
[8] Nigmatulin R I 1990 *Dynamics of Multiphase Media* (New York: Hemisphere)
[9] Mednikov E P 1965 *Acoustic Coagulation and Precipitation of Aerosols* (New York: Consultant Bureau)
[10] Gubaidullin D A and Osipov P P 2013 *J. of Engin. Physics and Thermophysics* **86**(1) 50
[11] Tukmakov A L 2009 *Acoustical Physics* **55**(3) 342

Journal of Physics: Conference Series **567** (2014) 012038 doi:10.1088/1742-6596/567/1/012038

Oscillations of a homogeneous gas and drift particles in an external wave field an open tube

L A Tkachenko[1,2] and R G Zaripov[1,2]

[1] Institute of Mechanics and Engineering, Kazan Science Center, Russian Academy of Sciences, 2/31 Lobachevsky str., Kazan 420111, Russia

[2] Kazan (Volga Region) Federal University, 18 Kremlyovskaya str., Kazan 420111, Russia

E-mail: luda_tkachenko@inbox.ru

Abstract. The dynamics of a particle and a homogeneous gas in the external wave field open tube by different values of amplitude of piston travel at the first eigenfrequencies in the shock-free mode are experimentally investigated. It is found the velocity distribution of the gas flow in the external the wave field open tube in different cross sections at various amplitudes of piston travel. The dependence of the coordinates of the particle along the axis the outside of the tube at on the time at resonance frequency are obtained. The comparison of velocity flow of the gas and particle drift in the external the wave field open tube is performed.

1. Introduction

At present the study of wave dynamics of dispersed media is an urgent [1]. Such media are often found in modern technology. Therefore, it is important to study the resonance oscillations media in tubes and their influence on the dispersion system, for example, aerosols (a mixture of gas with droplets or solid particles). In the theoretical investigation of nonlinear oscillations of gas in an open tube having significant difficulties caused by different effects of a nonlinear wave field at the exit cross section. An analysis of theoretical and experimental work is shown in detail in review [2]. Discontinuous oscillations of gas near nonlinear resonances were studied [3]. External wave field in the mode of formation of shock waves was investigated in [4, 5]. Various nonlinear effects in the propagation of waves in the tube with the flanges on end are detected [6]. In experiments [7] appearance at the open end tubes of pulsating jet and vortex rings near a linear of the resonance was found.

When modeling the coagulation and deposition of aerosols arising the problem of studying the motion of single particles in a nonlinear wave field of the tube. The results of experimental studies of the motion of particles with different physical and geometrical parameters at nonlinear oscillations of a homogeneous gas along the axis of the open tube near the subharmonic resonances in [8, 9] are given. In these papers the shock wave mode of oscillation of gas at large excitation amplitudes (of the order of 0.1-0.4 bar) is considered. The shock-free mode at low amplitudes of excitation (about 0.01 bar) when the loss to the walls become significant, has not been studied.

[1] To whom any correspondence should be addressed.

In this connection the aim of this work is to study the dynamics of particle and determination of the velocity distribution of the gas flow in the external the wave field open tube in different cross sections at various amplitudes of piston travel at the first eigenfrequencies in the shock-free mode.

2. Experiment

Experiments carried out on a previously used to study the coagulation and deposition of the aerosol [10], which was upgraded to study the oscillation of a homogeneous gas and particle motion in the external the wave field open tube. Longitudinal oscillations of the gas in a horizontal quartz tube length of 1.06 m and of inner diameter of 0.0365 m are created by a cylindrical piston with a flat surface. Piston is driven by means of the vibration generator TV51075 with a power amplifier VAA 120 firms TIRA. The measurement and control of the sinusoidal oscillations are done by the software module SineVIEW (VR610) installed on the computer by means of a piezoelectric IEPE accelerometer with integrated electronics 4513-001 firm Bruel & Kjaer and controller type VR8500-1 firm Vibration Research Corporation. The values of frequency and of amplitude of piston travel are set with an accuracy of better than 10^{-6} Hz and 10^{-7} m, respectively. Experiments were performed at gas oscillations near the first eigenfrequency $\nu_1 = 78.6$ Hz.

For measurement the gas flow velocity with an accurate to 0.01 m/s was used thermoanemometer ATT-1004 firm AKTAKOM with a thermistor placed in a small measuring sensory probe head on a telescopic handle. Telescopic handle attached to a traversing gear that allowed her to move along and perpendicular to the pulsating jet. Data from the hot-wire anemometer are supplied to the computer across RS-232, and are processed by a special program

For studying of drift of a particle along an axis of a tube the fishing line with a diameter of $0.3\cdot10^{-3}$ m was stretched. Fishing lines with tension was fixed on the one hand near the piston and the other hand on rack outside of the pipe at a distance of 0.3 m. The particle represented a washer from expanded polystyrene with a diameter of $0.9\cdot10^{-3}$ m and a thickness of $0.4\cdot10^{-3}$ m. The cylindrical polyethylene tubule $2.1\cdot10^{-3}$ m long with an internal diameter of 0.8·10-3 m and external diameter of $1.3\cdot10^{-3}$ m was inserted into an opening of a washer. The weight of the particles is $m = 4.6$ mg. Visualization of particle motion was carried with a digital SLR camera Canon EOS 650D with the lens Kit EF-S 18-55mm IS II Black. Particle was placed outside of the tube at a distance of 0.03 m from the open end and oscillations were started. For processing the results of the data transferred from the camera to the computer. Video of by means Virtual Dub 1.5.3 editor broken into frames and each frame correlated with the time from the start of shooting that allows fixing the position of a particle at different times.

3. Results

Figure 1 shows the dependence of the velocity of gas flow on the distance r from the tube axis for different distances from the open end of the tube. At small distances from the open end of the tube ($x = 5\cdot10^{-3}$ m) accurately expressed stream kernel with constant values of the gas velocity is observed. At a distance $x = 14\cdot10^{-3}$ m core flow practically disappears. And a smooth change in the distribution of the gas flow velocity to the value $\upsilon = 0.2$ m/s is observed that is connected with dispersion of a kernel. Further increase in distance leads to that the value of the velocity of the gas from maximum its value decreases to $\upsilon = 0.1$ m/s. Thus, the increase of the distance from the open end of the tube more than $14\cdot10^{-3}$ m leads to the fact that in the dependence velocity of gas on the radial coordinate constant values are not observed. Thus there are a smooth change of velocity values. And already at a distance $x = 52\cdot10^{-3}$ m are available only insignificant increase in speed of gas from $\upsilon = 0.1$ m/s. Comparison of the results with the mode with the formation of shock waves, in which the core flow is 1.5 to 6 radii of the tube indicates that the core flow in the shock-free mode is less than 0.7 the radius of the tube. This is due to the significant influence of the dissipative losses.

Figure 2 shows the values of the gas flow velocity to the tube axis on the amplitude of piston travel at different distances from the tube end. Obviously, increasing the amplitude of piston travel leads to an increase of the gas velocity for all distances from the open end of the tube. At the same time at a distance of $x = 5{\cdot}10^{-3}$ m from the end of the tube nonlinear character of the increase the gas flow velocity in 4.5 times from $\upsilon = 0.6$ m/s at an amplitude of $l = 0.25{\cdot}10^{-3}$ m to $\upsilon = 2.7$ m/s at amplitude $l = 1.75{\cdot}10^{-3}$ m is observed. Further increase in the distance leads to that an increase in velocity with increasing amplitude of piston travel becomes linear. Figure 3 shows the dependence of the flow velocity along the tube axis of the distance from the open end for different values of amplitude of piston travel. As seen from the figure, there is a sharp decrease in the velocity: for example, for $l = 1.75{\cdot}10^{-3}$ m increase in the distance by 10 times (from $x = 5{\cdot}10^{-3}$ m to $x = 52{\cdot}10^{-3}$ m from the open end of the tube) leads to reduction of the velocity of approximately by 25 times (from $\upsilon = 2.7$ m/s to $\upsilon = 0.1$ m/s). The area of the jet is practically not observed at a distance equal to the inner diameter of the tube, as opposed to the regime with the formation of shock waves, where this region occupies the to 5 dimensions of the internal diameter of the tube [11]. It is connected with expansion of a cylindrical jet during removal from an open end already at a distance of $x = 14{\cdot}10^{-3}$ m.

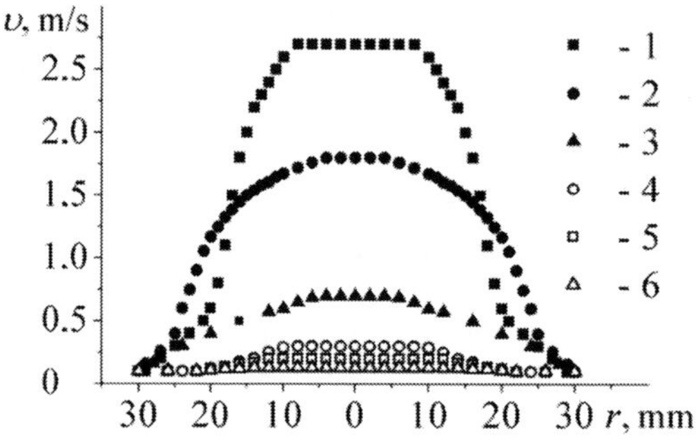

Figure 1. The dependence of the velocity of the gas flow on the distance from the axis of the tube at the amplitude of piston travel $l = 1.75{\cdot}10^{-3}$ m: 1 – $x = 5{\cdot}10^{-3}$ m, 2 – $x = 14{\cdot}10^{-3}$ m, 3 – $x = 28{\cdot}10^{-3}$ m, 4 – $x = 42{\cdot}10^{-3}$ m, 5 – $x = 45{\cdot}10$–*3* m, 6 – $x = 52{\cdot}10^{-3}$ m.

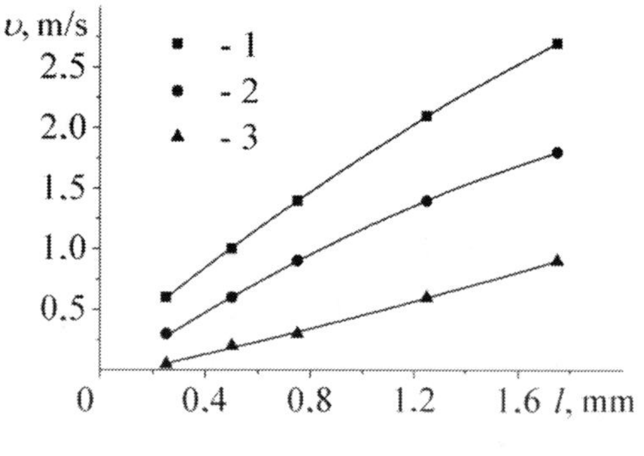

Figure 2. The dependence of velocity of the gas flow on the tube axis on the amplitude of piston travel: 1 – $x = 5{\cdot}10^{-3}$ m, 2 – $x = 14{\cdot}10^{-3}$ m, 3 – $x = 28{\cdot}10^{-3}$ m.

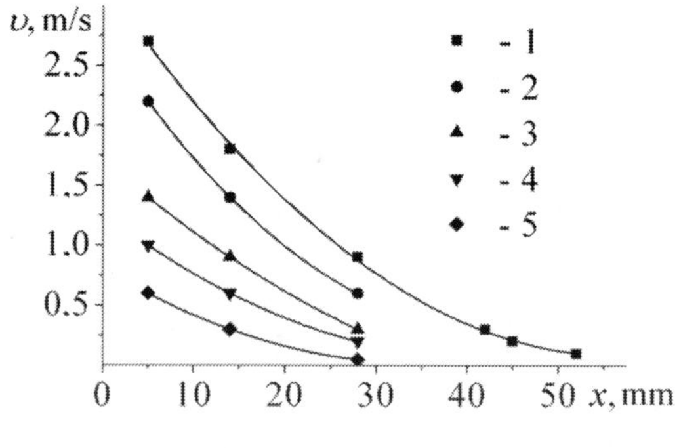

Figure 3. The dependence of velocity of the gas flow on the tube axis on the distance from the open end: 1 – $l = 1.75{\cdot}10^{-3}$ m, 2 – $l = 1.25{\cdot}10^{-3}$ m, 3 – $l = 0.75{\cdot}10^{-3}$ m, 4 – $l = 0.5{\cdot}10^{-3}$ m, 5 – $l = 0.25{\cdot}10^{-3}$ m.

Now consider the dynamics of the particle. In the external wave field of the tube significant oscillations of the particle was not observed. Figure 4 shows the dependence of the coordinates of the particle with $m = 4.6$ mg along the tube axis from the time at the resonance frequency of the amplitude of piston travel $l = 1.75 \cdot 10^{-3}$ m. The points – experimental data, the solid line – approximation. The data allows to determine the drift velocity of the particle from the open end towards the outside of the tube. This velocity was determined as the ratio of a given distance to time of the travel this distance. As a result, for a given amplitude of piston travel have the value $V_p = 0.36$ m/s. Measurement of the gas flow velocity in the external the wave field for the conditions of this experiment gave the value of $V = 0.8$ m/s. Unlike the velocity due to the presence of friction on the fishing line, and the influence of the hydrodynamic flow around the particle.

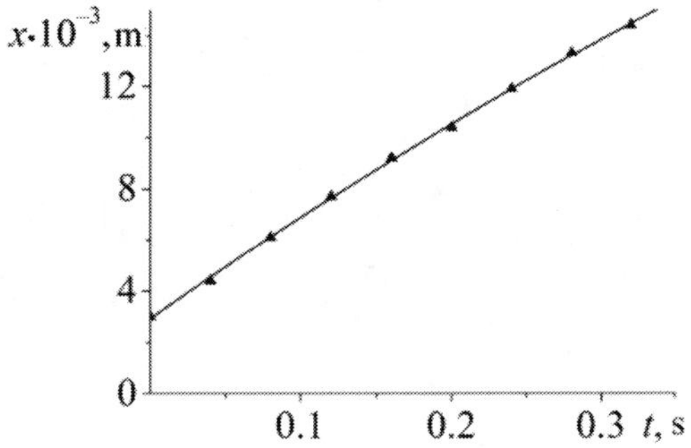

Figure 4. Dependence of the coordinate of the particle along the tube axis on the time near the open end outside of the tube for the amplitude of piston travel $l = 1.75 \cdot 10^{-3}$ m. The solid line - approximation.

4. Conclusions

Thus the dynamics of particles and of the gas flow in the external the wave field open tube at of amplitude of piston travel at the first eigenfrequencies in the shock-free mode are experimentally investigated. It was found that the core of the flow in the shock-free current is less than 0.7 the jet radii that is 2-9 times less than in the case of oscillations with the formation of shock waves. Increasing of amplitude of piston travel leads to an increase in the gas flow velocity for all distances from the open end of the tube. It is revealed that near the open end of the tube with increasing of amplitude of piston travel by 7 times velocity of the flow of gas in the resonance increases by 4.5 times, and increasing the distance by 10 times reduces the velocity of approximately 25 times. It is shown that near the open end outside the tube is an increase in the particle velocity with an increase in the piston displacement amplitude. The motion of the particle from the open end along the axis of the pipe takes place practically without hesitation. It is shown that near the open end outside the tube is an increase in the particle velocity with an increase in the amplitude of piston travel. The motion of the particle from the open end along the axis of the tube takes place practically without oscillations.

Acknowledgments
This work was funded by the subsidy of the Russian Government to support the Program of Competitive Growth of Kazan Federal University among World's Leading Academic Centers.

References

[1] Nigmatulin R I 1990 *Dynamics of Multiphase Media* (New York: Hemisphere)
[2] Ilgamov M A, Zaripov R G, Galiullin R G and Repin V B 1996 *Appl. Mech. Rev.* **49** 137
[3] Zaripov R G 1977 *Acoustical Physics* **23**(3) 378
[4] Repin V B, Novikov Yu N and Dement'ev A P 1990 *J. of Eng. Phys. and Thermophys.* **58**(6) 1027

[5] Galiullin R G and Khalimov G G 1979 *J. of Eng. Phys. and Thermophys* **37**(6) 1043
[6] Salikuddin M and Broun W H 1985 *J. of Sound and Vibr.* **106** 71
[7] Hudson G E 1955 *J. of Sound and Vibr.* **42**(3) 406
[8] Gubaidullin DA , Zaripov R G, Galiullin R G and Tkachenko L A 2007 *J. of Probl. of Power Eng.* (9-10) 3
[9] Gubaidullin DA , Zaripov R G, Galiullin R G and Tkachenko L A 2010 *J. of Probl. of Power Eng.* (9-10) 15
[10] Gubaidullin D A, Zaripov R G and Tkachenko L A 2013 *Doklady Physics* **58** 392
[11] Galiullin R G, Revva I P and Khalimov G G 1982 *Acoustical Physics* **28**(5) 617

Journal of Physics: Conference Series **567** (2014) 012039 doi:10.1088/1742-6596/567/1/012039

Use of compact spectrometer for plasma emission qualitative analysis

A A Uhov, V A Gerasimov, D K Kostrin, L M Selivanov

Faculty of Electronics, Saint Petersburg Electrotechnical University "LETI", Saint Petersburg, Prof. Popov str. 5, 197376, Russia

E-mail: aauhov@yandex.ru, vl.gerasimov@mail.ru, dkkostrin@mail.ru, lmselivanov@gmail.com

Abstract. Usage of compact optical spectrometer for plasma emission analysis is considered. Capabilities of specially developed personal computer software to identify specific emission spectral lines are shown. Difficulties of analysis of complex plasma compositions emission spectrums are described.

Plasma analysis by means of spectral methods is widely and successfully used since years [1-5]. To determine plasma gas composition, both qualitatively and quantitatively, plasma emission spectrum is to be analyzed. If the technological process to be analyzed is not yet well-elaborated, good choice of equipment for spectral analysis is some compact general-purpose optical spectrometer such as ISM3600 developed by St. Petersburg Electrotechnical University [6].

PC software used to control the spectrometer and to visualize and analyze the spectral data obtained from ISM3600 is called Aspect2010 [7]. One of main functions of this software is plasma emission spectrum analysis. For plasma composition manual analysis it is useful to display some spectral lines of certain periodic table element on chart (figure 1).

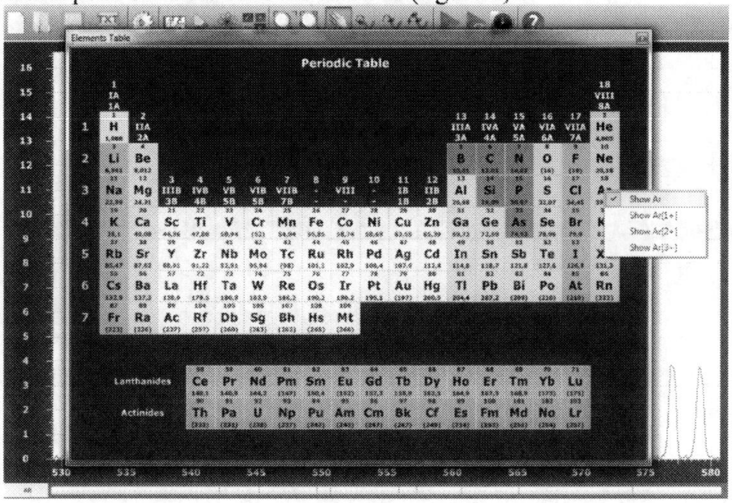

Figure 1. Periodic table element selection.

Journal of Physics: Conference Series **567** (2014) 012039 doi:10.1088/1742-6596/567/1/012039

Spectral lines corresponding to selected periodic table elements are displayed on the same chart together with measured spectrum that allows detecting of such elements presence in plasma composition [8]. Figure 2 shows Aspect2010 chart containing measured spectrum and spectral lines of Ar which was selected from periodic table window.

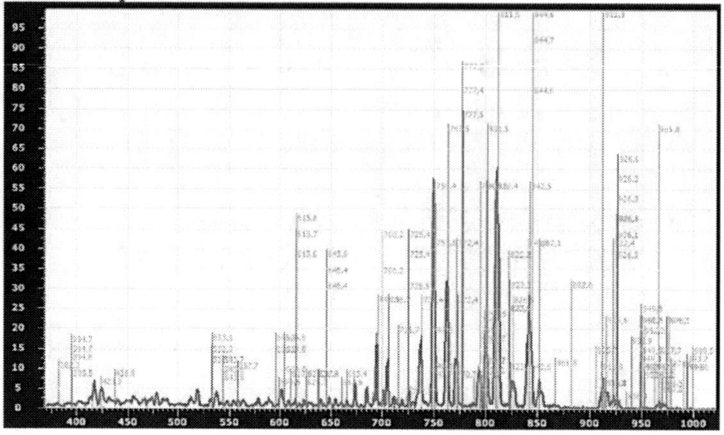

Figure 2. Plasma spectrum with spectral lines markers.

Usually technological chamber contains more than one periodic table element so software can display few elements spectral lines simultaneously. Resolution of ISM3600 spectrometer is approximately 1.5 nm so lines located close to each other on wavelength axis can visually join on chart. Such lines can belong to different periodic table elements that create additional difficulties for analysis of plasma composition. Figure 3 depicts fragment of spectrum containing oxygen and argon spectral lines. 798-802 nm wavelength range contains superposition of not resolved by spectrometer lines of the two elements.

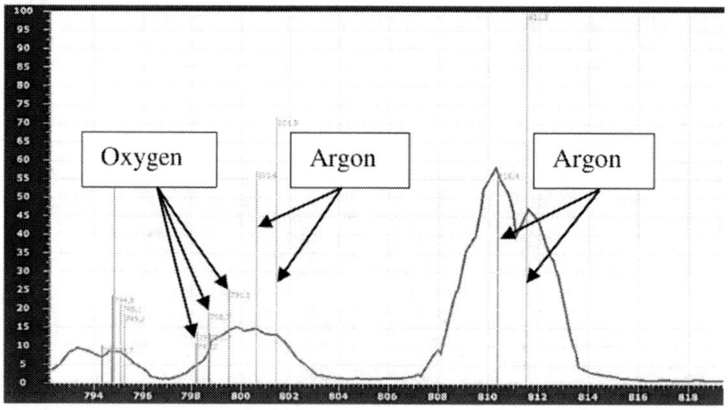

Figure 3. Spectrum with oxygen and argon spectral lines

Components utilized in technological processes typically have complex spectrums with many spectral lines and usually it is possible to find wavelength band where absence or presents of certain element can be easily detected. Example of such condition is shown in figure 4: spectrum (1) contains no tantalum while spectrum (2) with tantalum present is obviously different.

Software also allows comparison of two spectral lines energies that handy for simple emission analysis. Parameters for this function entered by user are two wavelength bands. User defines these bands manually in such a way so line of interest suits its band. Software integrates spectrum data within both specified bands. Each band must not be narrower that spectral resolution of ISM3600 (1-2 nm) and estimated width of real spectral line. When set up, this software function integrates data within both bands (Σ1, Σ2) and calculates ratio of these values (Σ1/Σ2). Results of calculations are displayed on top of main spectrum plot as set of values and small time chart of (Σ1/Σ2). This chart represents the plasma composition dynamics in a descriptive visual way. Chart window is fixed to display last 190 measured and calculated data points.

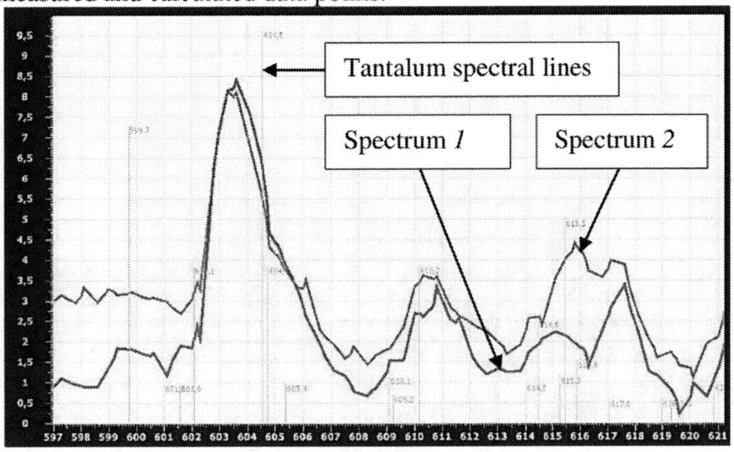

Figure 4. Plasma spectrums with and without tantalum present

Special analytic module of software can be used to process the experimental spectral data. Interface of this module is shown in figure 5.

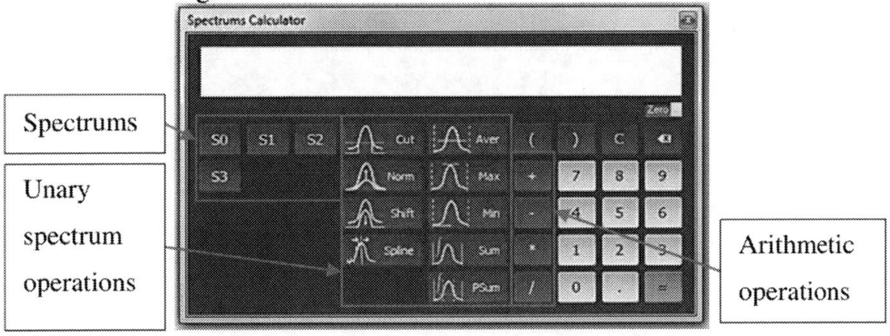

Figure 5. Spectrums calculator user interface

Buttons S0-S3 shown in figure 5 correspond to four spectrums on the chart. Mouse pointer move over certain button highlights its curve on chart that allows easy selection of proper spectrum. When some spectrum is selected it is possible to process it using some function (see "Unary spectrum operations" group in figure 5).

List of available operations applied to one spectrum:

- reverse-clamping to the specified level (values below level are zeroed);
- normalization to 100%;
- shift of spectrum down (negative values are zeroed);
- spectrum spline filtration;
- the spectral line half-width calculation in specified spectrum band;
- maximal value calculation in the specified spectrum band;
- minimal value calculation in the specified spectrum band;
- calculation of the area under spectral curve in specified spectrum band;
- peak weight calculation in specified spectrum band (peak is automatically detected, then area under peak is calculated).

There is possibility to use some binary operation such as subtraction or multiplication in case two spectrums have been selected. Set of available options allows creation of quite complex formulas. For instance, comparison of energies of two spectral peaks requires not only simple comparison of their magnitudes. First of all it is necessary to calculate areas covered by these peaks and then – ratio of these areas. Result of each operation can become a part of more complex formula.

Set of features for spectral emission analysis offered by Aspect2010 software is quite reach but of course these features are suitable for some manual analysis for technology tuning and not intended to be used in any kind of automatic process control.

References

[1] Ukhov A A, Gerasimov V A and Kostrin D K 2013 *Izv. SPbGETU LETI*, **8**, 10-14
[2] Bystrov U A, Kostrin D K and Lisenkov A A 2014 *Vac. Tech. and Technol.*, **1/14**, 164-66
[3] Kolgin E A, Ukhov A A and Savushkin A V 2008 *Peter. Zh. Electron.*, **2-3**, 120-26
[4] Komlev A A, Komlev A E and Ukhov A A 2012 *Vac. Tech. and Technol.*, **4/22**, 245-48
[5] Kostrin D K 2013 *Izv. SPbGETU LETI*, **7**, 15-19
[6] Kostrin D K and Ukhov A A 2013 *Biotechnosphera*, **3**, 21-25
[7] Yudin R V, Kostrin D K, Shishov D I and Ukhov A A 2013 *Izv. SPbGETU LETI*, **3**, 8-13
[8] Kostrin D K and Ukhov A A 2013 *Control Diagnosis*, **6**, 26-28

Phase and structural conditions of low-temperature plasma interaction products with steel

R A Valiev[1], Yu I Shakirov, A N Iliukhin

Branch of Kazan Federal University in Naberezhnye Chelny, 68/19 Mira Street, Naberezhnye Chelny, 423810, Russia

E-mail: rustvali@mail.ru

Abstract. Structural-phase research of low-temperature plasma interaction products with carbonaceous steel is conducted in the discharge between solid electrode and water. Such a discharge at a certain electric parameters leads to a powder formation of spherical particles with a diameter 10^{-6} – 10^{-3} m. A scale on surfaces of a steel electrode and the powder synthesized from it are subjected to a comparative analysis. Qualitative and quantitative phase characteristics of these products are different and depend on conditions of their formation. Their basic phase components are various ferric oxides: magnetite Fe_3O_4, wustite FeO, hematite α-Fe_2O_3. Magnetite is contained in the powder synthesized at normal atmospheric conditions in the greatest quantity. Atmospheric pressure reduction and discharge implementation in a nitrogenous aerosphere considerably reduce quantity of magnetite in a formed powder. Diffusion speed in oxide layers and reaction temperature also influence electrode iron oxidation process. The formation mechanism of ferritic powder from steel under the influence of discharge plasma with a liquid electrode is offered on the basis of the gained results.

1. Introduction

The processes occurring at affecting on electric discharge plasma substance with use of liquids are the least known. The particular interest represents a case when solid electrode is taken out from an electrolyte, and the discharge occurs between solid and liquid electrodes [1]. Though recently there were the works devoted to studying of electrophysical characteristics of the similar discharge between metal and an electrolyte [2, 3, 4], plasma interaction with a solid body in the presence of erosion is a little investigated. Steel application as plasma machining object and its property modification is one of discharge use directions with a liquid electrode [5, 6]. The great value for physical and chemical processes understanding of low-temperature plasma interaction with a solid body surface has a structure and phase research of products formed.

2. Experiment

The discharge is carried out between an electrolyte and a solid electrode at constant voltage 1,3 kW and a current up to 1 A. A process water served as an electrolyte, and carbon steel is used as a solid electrode with a carbon content from 0,1 to 1,0 % and a cross-section diameter from 2 to 10 mm.

The structure and phase condition of solid products of plasma interaction with iron-carbon alloy is defined by methods of X-ray structure analysis and Mossbauer spectroscopy. X-ray phase analysis was

[1] To whom any correspondence should be addressed.

based on difference of interplanar spacing and various phases lattice constant, Mossbauer spectroscopy is of isomer shift, quadrupole splitting and magnetic fields on iron cores. Application of Mossbauer Effect in various geometry allowed gaining the information, as by volume, and on a surface of investigated materials.

With a research objective of occurring processes in development, especially on a discharge initial stage, interaction product analysis was carried out stage by stage - depending on duration of discharge.

3. Results

At the very beginning of electric discharge there is visually defined layer on a narrow circle of its basis on a face surface of an electrode steel, a bit later - the first red-brown stains. In some seconds all face surface of steel becomes covered by a thin layer of rust. Then the electrode end heats up too much, the discharge occurs from all face surface of a steel, and it is covered by the black color layer which thickness gradually increases. Further there is a hemispherical drop on the electrode end, and its melt is spluttered in a liquid electrode where particles are intensively chilled and solidify in the spherical form, forming a powder.

The powder gained at current density on an electrode up to 10 A/cm2, consists of black spherical particles with a diameter from 1 micron to 1 mm. Further increase in density of a current leads to instability of dispersion process and formation of balls with a diameter 1 – 2 mm. At current density above 20 A/cm2 powder synthesis is stopped and the hollow particles are formed which transverse size reaches 5 mm.

The basic phase components of plasma interaction products with steel are various ferric oxides: magnetite Fe_3O_4, wustite FeO, hematite α-Fe_2O_3. However qualitative and quantitative phase compositions of these products (a scale on a steel electrode surface, powder and balls) are different and depend on conditions of their formation in an electric discharge with a liquid electrode. At the same time no ferric oxide gamma modification – maghemite γ-Fe_2O_3 is detected in each of them that follow from precision measurements of spinel structure and Mossbauer spectrum lattice spacing.

Originally appeared on a lateral surface of electrode steel the red-brown component is hematite α-Fe_2O_3. The scale formed on an electrode at much discharge times, basically consists of magnetite Fe_3O_4 and hematite α-Fe_2O_3 though it contains wustite FeO in insignificant quantity.

A dispersion powder is a mix of two ferric oxides Fe_3O_4 and FeO, though the magnetite phase considerably prevails and makes 80 – 90 % of volume. At the same time, as Mossbauer conversion spectra interpretation shows, the powder particles surface layer (about 0,1 μm) mainly consists of hematite α-Fe_2O_3.

The magnetite maintenance in a powder depends on many factors, including a discharge current, current density, carbon concentration in a steel electrode, and also a size of particles. At atmospheric pressure reduction the quantity of magnetite in a powder decreases and accordingly the wustite content increases. The same situation is also observed at synthesizing of particles in a nitrogenous aerosphere.

Precision measurements showed a little downgraded value of powder magnetite lattice parameter in comparison with stoichiometric Fe_3O_4. Intensiveness lines relation in synthesized particles magnetite Mossbauer spectrum that corresponds to ferric ions in tetrahedral and octahedral lattice points, also exceeds value, typical for a stoichiometric composition Fe_3O_4 [7]. The considerable broadening of FeO lines on diffractograms in comparison with equilibrium wustite is additionally revealed.

At high current densities metal balls are formed, and a main phase in them is, as in electrode steel itself, a solid solution on the basis of an alpha iron. Though the initial electrode steel had the equilibrium structure containing ferrite (a carbon solid solution in α-Fe) and the insignificant quantity of cementite (Fe_3C chemical compound), in composition of the above mentioned balls, except α-Fe, an austenite is available (a carbon solid solution in γ-Fe). The surface of these balls is covered by oxide crust which has phase composition qualitatively similar to a powder: Fe_3O_4 and FeO.

4. Discussion

Originally appeared on a surface of an electrode steel poorly reinforced layer on a narrow circle of its basis is a result of its heat treatment. However the discharge between a steel electrode and process water at the accepted modes practically does not allow reinforcing steel thermally: high temperatures developing in plasma promote further active interacting of ions being in excited state and electrons with formation of various compounds of iron with oxygen.

Oxidation process consists not only in a chemical compound of oxygen and iron, but also in diffusion of their atoms through the multiphase oxygenated layer. As the iron oxidizing reaction goes faster than diffusion through a scale layer, so at dense oxide film a speed of a scale build-up is defined by diffusion speed of atoms through its thickness that in turn depends on temperature and a oxide film structure. Magnetite and hematite have difficult cubic and rhombohedric crystal structure. Wustite crystallizes as a penetration phase, and it has vacant sites on a place of missing ions of iron that causes the big factor of diffusion in it. Thus it is necessary to mean that wustite formation in comparison with other ferric oxides is favorable thermodynamically only at higher temperatures [8].

At second discharge time the energy density on a face surface of an electrode is strongly non-uniform, and almost all energy is necessary on an external circle of the basis. Therefore the temperature of all surfaces of steel poorly raises, these can explain the absence of wustite in the oxygenated film at little discharge time. At increase in duration of plasma action the steel strongly heats up, the discharge is supported by all face surface of an electrode, and it oxidizes at such temperatures, forming Fe_3O_4 and FeO. The scale layer build-up on a steel surface is sharply sped up because of formation loose oxide FeO instead of dense oxide Fe_3O_4, and α-Fe_2O_3. The different concentration relationship of iron and oxygen atoms on the depth, caused by limiting diffusion, leads to decrease of magnetite content in a scale on depth.

The temperature, apparently, is a major factor causing difference in phase compositions of a powder and a scale on a surface of an electrode. Oxide ratio share in scale structure depends on temperature at which there is an oxidation of iron [9]. A characteristic point is eutectoid temperature (570°C under equilibrium conditions) below which the oxygenated layer consists only of hematite and magnetite, and above - among all three oxide types, however thus the hematite α-Fe_2O_3 content in comparison with magnetite Fe_3O_4 and wustite FeO is not enough. Indirectly this fact allows assuming range of temperature values at which there are processes of powder formation and the oxygenated layer on an electrode under the plasma action.

The additional factor of magnetite quantity raise in an oxide mixture, formed at plasma action on steel in the discharge with water, is ability of iron to react directly with its steams with Fe_3O_4 formation [10].

Oxide ratio share in the gained product is largely defined by concentration of oxygen in the dispersion environment. It decrease quantity of magnetite and accordingly increases the content of poorer oxygen wustite phase at atmospheric pressure reduction and at powder synthesizing in a nitrogenous atmosphere. The oxygen deficiency also leads to lack of hematite in powder particles.

The concentrated energy flows create special conditions of product synthesizing that explains formation in the discharge of non-equilibrium structures. A deviation of magnetite and wustite content from equilibrium, vacancies in their crystal lattices are developing process of these plasma interaction features with substance. Formation of unstable austenite in metal balls also explains a high speed cooling of steel particles in water (liquid electrode) from a liquid state.

The gained results allow to present synthesizing process of ferritic powder from steel under discharge plasma action with a liquid electrode. High-energy plasma action on a steel leads to formation on its face surface of a hemispherical drop of molten metal. High temperatures and water vapors cause presence of a magnetite and wustite mixture on an external surface of this drop with predominance of the first. The inner shell represents the molten metal kept by a surface tension force on the end of an electrode. When gaseous impurities burst out, they carry away material parts. A surface tension force gives them the spherical form, and extra fast cooling in a liquid electrode fixes their form.

Because the body-section phase composition of a drop is defined by diffusion of atoms at iron oxidation (magnetite predominates on a surface, deeper is wustite, metal is inside) its phase composition depends on volume of a material carried away by a particle. Oxidation proceeds before full cooling of a particle, however it already occurs at lower temperatures when thermodynamically favorable is α-Fe_2O_3 formation, owing to that the particle surface becomes covered by a thin hematite layer.

5. Conclusions

At constant voltage above 1 kW and a current up to 1 A the electric discharge between a steel electrode and process water leads to formation of a spherical particles powder with sizes from micron to millimetric values. Sphericity of particles is provided with a surface tension force and high speeds of their cooling in a liquid electrode.

Process is accompanied by iron oxidation and its oxide Fe_3O_4, FeO and α-Fe_2O_3 formation, except γ-Fe_2O_3. Major factors defining qualitative and quantitative phase content of plasma interaction products with steel are the temperature of their formation and diffusion speed of their elements.

Iron oxidation level decreases from a surface to the centre of particles of the synthesized powder, thus their surface is covered by a thin α-Fe_2O_3 layer. Its quantity practically does not influence volume phase powder content which basically consists of Fe_3O_4 and FeO. A prevailing component of this mixture is Fe_3O_4 which content decreases at reduction of oxygen concentration in discharge environment.

The high substance content in a composition of a powder including into ferrite group with spinel structure, represents practical interest for magnetic materials powder metallurgy. Special value thus has magnetite powder particle form sphericity which can be an important condition of isotropy of its magnetic properties.

References

[1] Gaisin F M, Son E E, Shakirov Yu I 1990 *Volume discharge in the vapor-gas medium between solid and liquid electrodes* (Moscow: VZPI Press) p 88
[2] Miftakhov M N, Tazmeev Kh K, Tazmeev A Kh and Fridland S V 2006 *J. of Engin. Phys. and Thermophys.* **79** 532
[3] Tazmeev K K, Arslanov I M and Tazmeev G Kh 2013 *Appl. Phys.* **4** 33
[4] Kashapov L N, Kashapov N F and Kashapov R N 2013 Research of plasma-electrolytic processes in various ratio of the area anode to the cathode *5th Rep. Scien. Tech. Conf. on Low-Temperature Plasma During the Deposition of Functional Coatings (Kazan; Russian Federation; 4-7 November 2013) J. of Phys.: Conf. Ser.* **479** 012005
[5] Kashapov L N, Kashapov N F and Kashapov R N 2013 Investigation of the influence of plasma-electrolytic processing on the surface of austenitic chromium-nickel steels *5th Republican Scientific Technical Conference on Low-Temperature Plasma During the Deposition of Functional Coatings (Kazan; Russian Federation; 4-7 November 2013) J. of Phys.: Conf. Ser.* **479** 012003
[6] Valiev R A, Gaisin F M and Shakirov Yu I 1991 *Soviet Powder Metallurgy and Metal Ceramics* **30** 448
[7] Rubinstein M and Forester D W 1971 Investigation of the insulating phase of magnetite by NMR and Mossbauer effect *Solid State Commun.* **9** 1675
[8] *Physico-chemical properties of oxides: A Handbook* 1978 (Moscow: Metallurgy) p 471
[9] Umansky Y O, Ckakov Y A *The Physics of Metals* 1978 (Moscow: Atomizdat) p 352
[10] Tamarov M A *Inorganic chemistry* 1994 1978 (Moscow: Medicine) p 480

Studying optical characteristics of an axial symmetric non stationary plasma

R S Toukhvatoulline, F F Zalyalieva, T A Ashrapov and F I Bashirov
Department of Physics, Kazan National Research Technical University named after A N Tupolev, 10, K.Marx Str., Kazan, Republic of Tatarstan, 420111, Russia

E-mail: yafidka-kai@mail.ru

Abstract. The distribution function of arc pulsations was obtained as functions of current intensity, gas flow rate, and distance from the cathode in a plasmatron with a fixed arc length. The intensities of plasma emission were measured by the method of transverse images. As the results, the temperature distribution over the arc radius was determined and the effect of pulsations on the measurement precision was noted.

1. Introduction

A significant divergence of the results on the temperature distribution in electric arc and microwave plasmas occurs by their spectroscopic studies often. Partially, it is connected to the fact that the common spectroscopic methods provide reliable results when the plasma properties are stable on time. If the plasma is non stationary, the distribution of the investigated quantity will be determined with a certain error, which depends on the degree of plasma instability and the kind of time averaging of the quantity.

The present work is aimed at experimental study of pulsations in axial symmetrical, low temperature plasma accounting for the effect of these pulsations on the temperature of plasma.

Apparatus and experimental procedures.

The plasmatron under experiment was composed of the sectioned interelectrode insert and the vortex gas stabilizator of arc. The arc chamber diameter was equal to d = 1 cm and the length of interelectrode insert was a = 10 cm. The plasmatron had included also the cathode and anode units, the interelectrode insert sections, the gas curling ring, insulating and sealing spacers, flanges, bolts, and nuts required to assembling the plasmatron. The anode and the interelectrode insert sections were manufactured of copper. A hafnium rod embedded into the copper holder was the cathode. The insert sections were insulated from each other by Teflon spacers and were cooled by water. A sections being the optical one had a rectangular slit with 1 cm of height and 0.1 cm of width. The slit was closed by K8 optical glass. This section makes it possible to observe the arc column in a particular channel cross section. By varying the optical section location, it makes possible to investigate the diver arc domains. The arc was formed between the cathode and the cylindrical anode. The air was fed tangentially into the arc chamber through the holes of the curling ring of the initial cross section. To provide laminar gas flow, the internal plasmatron channel surface was polished.

A NAC Memrecam GX8 high-speed camera was used to record the arc pulsations. The emission of selected arc cross section was projected on the camera objective. The camera provides made recording at a highest sensitivity (> 20000 ISO) with the minimal exposure time of 600 ns. During the experiment, the digital image of the arc position was registered in the camera's memory to be afterward transferred for processing to a PC. Using specific image processing software, we was obtaining the registrogram of the arc pulsations in the plasmatron channel.

Results

We investigated the arc pulsations within a continuous current range I = 100 – 250 A, the gas flow rate 0.2 – 8 g/s, and the distance from the cathode z = 1.8 – 8.7 cm. It was detected that the arc undergoes low frequency (~300 Hz) pulsations well as high frequency (~104 Hz) ones and complex physical processes take place in the arc chamber. We studied the high frequency arc pulsations at the current intensities I = 100, 150, 200, and 250 A, the gas flow rates G = 0.4 – 8 g/s, and the distances from the cathode z = 1.8 – 8.7 cm.

It follows from the registrogram analising at low gas flow rates that the pulsations are random and submit the normal distribution law

$$\varphi(x) = \frac{1}{\sigma_x\sqrt{2\pi}} e^{-x^2/2\sigma_x^2} \qquad (1)$$

The registrogram processing makes it possible to determine the standard deviations of the arc axis σ_x depending on the current intensity, the gas flow rate, and the distance from the cathode. The analisis of the experimental results shows that the pulsation amplitude growth with the increase of G and z and with the decrease of I. The σ_x values obtained here are in a good agreement with the results of [1]. Thus, we be able to constat that, at low rates of gas laminar flow, the arc pulsations may be described by the normal distribution law (1). The random pulsations of the arc are caused by several factors such as turbulence in the gas flow, the arc interaction with its own magnetic field and the channel wall, rotation of the preanode section of arc positive column, and the smallscale shunting of discharge [2]. At a flow rate of more than 2 g/s, shape of oscillations has another form (Figure 1).

It is seen from Fig. 1 that harmonic oacillations superimpose on random fluctuations. Distribution function of oscillations of the arc axis has been constructed with accordance to the records. This results in the axis of arc oscilates in the channel region of 2.5 mm and the most probable position of the arc is not placed on the axis of the channel, but at a some distance from it and, in addition, with increasing the current intensity, the amplitude of oscillation increases. Taking into account these experiments one can propose the following model of an axial symmetric plasma arc: at low gas flow rates (up to 2 g/s), the axis of the arc makes random fluctuations obeying Gaussian low. At gas flow rates of more than 2 g/s, the arc begins to rotate as a whole (Fig. 2). The axis of the arc O' moves along a circumference of radius a in the plane z = const, the center of which is a subject of random fluctuations described by the formula (1). In this figure, the solid circumference curve represents the true size of the plasma, large dotted circumference – the observed size of the plasma and the small dotted circumference of radius a - track of rotation of the plasma of radius R. By the letters HOY, the laboratory reference frame is denoted and by H'O'Y' – the coordinate system associated with the plasma.

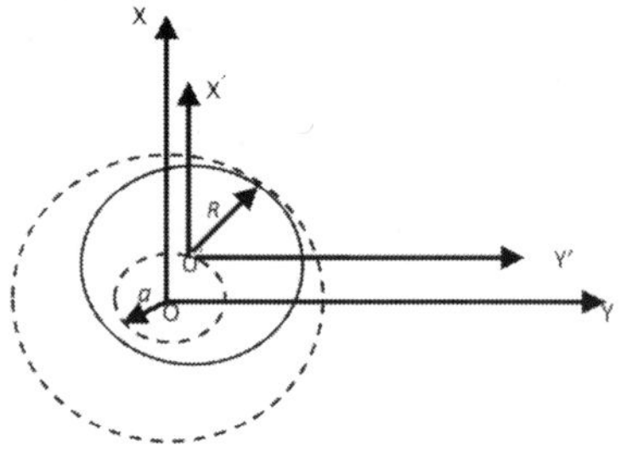

Figure 1. Schema of the pulsations of the arc plasma. XOY – the laboratory reference frame X'O'Y' – the coordinate system associated with the plasma. R – radius of the plasma, and a – the radius of the arc rotation.

The distribution function of harmonic oscillations is represented as:

$$\varphi_2(x) = \frac{1}{\pi\sqrt{a^2 - x^2}}, \quad a \le x \le a, \ \varphi_2 = 0, \ a < |x|. \qquad (2)$$

Consequently, the distribution function based on the random harmonic vibrations of O' is equal to:

$$\varphi(x)=\int_{-a}^{a}\frac{1}{\pi\sigma\sqrt{2\pi(a^2-\xi^2)}}\exp[-\frac{(x-\xi)^2}{2\sigma^2}]d\xi. \quad (3)$$

To determine the plasma temperature accounting for its spatial pulsations, it is necessary to obtaine the time averaged distribution of the emission intensity $I_\nu(x)$ over the height of the spectral line. Knowing the distribution function $\varphi(x)$, the proper distribution $q_\nu(x)$ can be determined by solving the Fredholm equation of the first kind

$$I(x)=\int_{-R}^{R}q(x')\phi(x-x')dx', \quad (4)$$

where R is the plasma radius. Here, the distribution function $\varphi(x\text{–}x')$ is described by the normal distribution law (1). The equation (4) has been solved by the Phillips–Toumi regularization method [12, 13]. We wrote the program in the C++ language. Figure 2 shows the results for the currents intensities 100 and 250 A, the gas flow rate 6 g/s, and z = 5.2 cm.

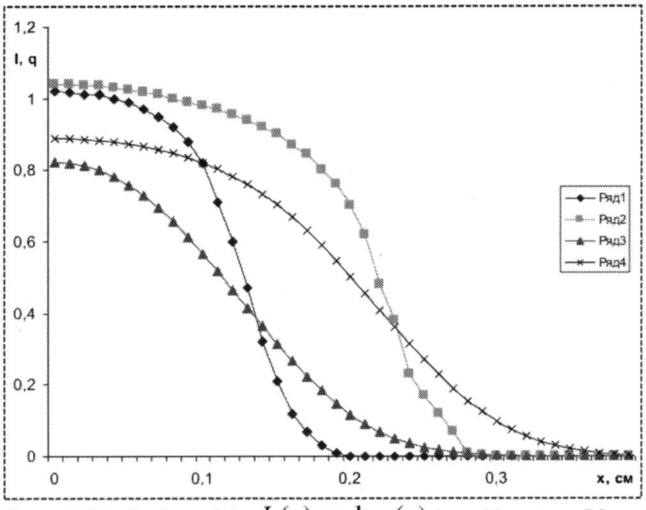

Figure 2. Distributions of the $I_\nu(x)$ and $q_\nu(x)$ intensities at z = 5.2 cm, G = 6 g/s at I = 100 A (1;3) and 250 A (2;4).

The $I_\nu(x)$ distributions were spectroscopically determined according to the NI 493.5 nm line; the $q_\nu(x)$ are found according to formula (4) by using of the experimental data on the $I_\nu(x)$ and distribution function (3). It is seen that the pulsations cause underestimation of the intensity at the arc axis and its overestimation at the periphery. With the current intensity increasing, the line profile becomes tighter.

From the measured values of $I_\nu(x)$ and $q_\nu(x)$, we found the radial emissivities and the temperatures. Figure 3 shows the results. We can see the electric arc pulsations influence on the character of the temperature distribution over the radius. Neglecting the arc pulsations results in temperature underestimation at the axis zone and its overestimation at the periphery, as well as in the temperature profile distortion.

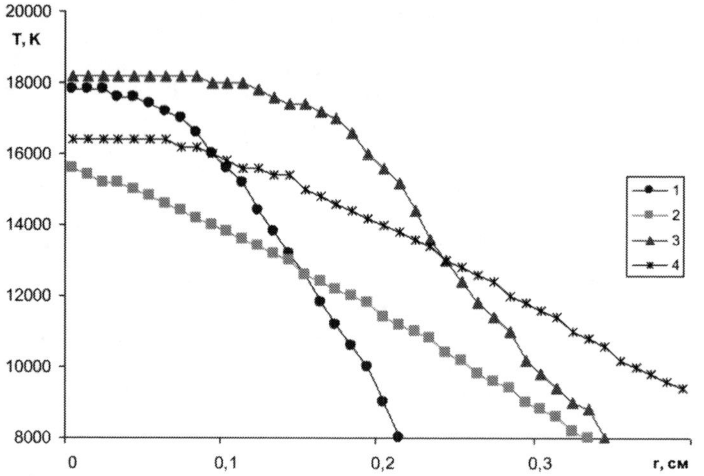

Figure 3. Temperature distributions over the arc radius at z = 5.2 cm, G = 6 g/s, I = 100 A (1;2) and 250 A (3;4).

References

[1] Zakirov I.M., Zalyalieva F.F., Timerkaeva D.B., and Tukhvatullin R.S., *High Temp.*, 2011, **vol. 49**, no. 3, p. 330.

[2] Dautov G.Yu., in *Modelirovanie i metody rascheta fiziko-khimicheskikh protsessov v izkotemperaturnoi plazme. Sbornik* (Simulation and Methods of the Calculation of Physical-Chemical Processes in Low-Temperature Plasma: A Collection of Papers), Moscow: Nauka, 1974, p. 185-208.

Journal of Physics: Conference Series **567** (2014) 012042 doi:10.1088/1742-6596/567/1/012042

Drift of a flat particle at longitudinal oscillations of gas in an open tube

R G Zaripov[1,2], L A Tkachenko[1,2] and Shaydullin L R[2]

[1] Institute of Mechanics and Engineering, Kazan Science Center, Russian Academy of Sciences, 2/31 Lobachevsky str., Kazan 420111, Russia
[2] Kazan (Volga Region) Federal University, 18 Kremlyovskaya str., Kazan 420111, Russia

E-mail: zaripov@imm.knc.ru, luda_tkachenko@inbox.ru, liqn@mail.ru

Abstract. The particle motion is experimentally investigated at nonlinear oscillations of gas in the tube and in the external field near the open end in the shock-free mode. Dependence is obtained for the coordinates of the particle along the tube from time for various frequencies and amplitudes of displacement of the piston. Drift is set for a particle from the open end of the tube to the piston. It is determined that the particle moves into an external field at the open end outside the tube without appreciable oscillations. Detected position of the particle at the open end, wherein the particle does not perform drift towards the inside and outside of the tube.

1. Introduction

Research of wave processes represents one of actual problems in restricted actuation environments. Studying of oscillations of environments with strongly non-linear fronts of waves of pressure has important applied value and their impacts on disperse systems which examples are gas mixes with drops or solids. The main mathematical models of wave dynamics of multiphase environments and theoretical and also experimental results are explicitly in [1]. Oscillating processes are studied in two-phase restricted environments in [2]. In work [3] the currents arising at acoustic resonance oscillations of gas in a tube are considered. Currents are considered in work [3], arising at acoustic resonance oscillations of gas in a tube. Effect of Prandtl number and absorption on the walls was investigated on the velocity of flow in viscous and heat conduction medium. Expressions are obtained for the longitudinal and transverse components of the flow velocity. Features are analyzed a drift of particles at longitudinal oscillations of gas in an open tube in work [4]. Behavior was studied spherical particles with various geometrical, physical properties along a tube axis at oscillations of gas in a tube in work [5]. Pressure oscillations were amplitude of 0.1 bar and above with the occurrence of periodic shock waves at the resonance frequencies in these studies. The review is provided theoretical and experimental works on non-linear oscillations with such amplitudes in [6]. Results of experimental studies are not available for the dynamics of the particle at longitudinal oscillations of gas in an open tube at the first natural frequency in the unstressed-wave mode at low amplitudes of excitation with the wave pressure of 0.01 bar up to the present time, which is necessary for a full analysis of the behavior of the particle. The need arises of studying the process of the particle drift over time in

[1] To whom any correspondence should be addressed.

different sections of the tube and the external wave field at different frequencies and amplitudes of excitation of the gas in the shock-free-wave regime.

2. Experimental setup

Researches are conducted on a modified setup which is earlier used for studying of the coagulation and deposition of an aerosol [7]. Scheme of setup is submitted in Figure 1. The glass tube 5 which is horizontally focused with a length of 1.06 m with internal diameter of 0.0365 m and external diameter 0.041 m, and the s vibration generator 12 of the TV51075 firm TIRA are given to horizontal situation. Vibration generator is fixed on a metal stand 19 which to build sustainable structures to vibration loads the ballast. Also, the machine bed is fixed to the base by means of brackets 22 and 10 M8, which through the nozzle 8 piston installed in the tube 5, which was stretched along the axis of the fishing line 7 with the particle 6 and mounted on a rack 1. The flat piston 11 creates exaltations of gas and has diameter of $d = 36.5$ mm which it moves under the sine law in the cylinder 21 from a two-stroke micromotor of MDS 218KU-K of internal combustion. The total weight of the piston with the system it is installed on the shaker table 150 g. The tube is held in a horizontal position by means of tie 4 fixed to machine frame 3. The end of the tube passes through the nozzle 2 which it was wrapped on the bench 3 fixed on a support material 26.

The opening is drilled in a nozzle at the piston by diameter of M5 for installation of a core 15. The rod has a length of 13 mm and M5 threaded holes along the axis in which the medical needle 16 is threaded and secured with a diameter of 0.8 mm. The end of a needle was on a tube axis. Fishing line 18 was extended to 0.3 mm in diameter through the needle. It was fixed with a tension on a rack of 1 (fig. 1) out of a tube apart of 300 mm with orientation along a tube axis. The nozzle was similar to the open end of the tube nozzle in the piston. Appearance of the particle is a washer of expanded polystyrene diameter of 0.9 mm and a thickness of 0.4 mm. The tubule was inserted into a washer opening it was cylindrical polyethylene with a length of 2.1 mm with a caliber of 0.8 mm and external 1.3 mm. Weight of a particle made 4.6 mg. The tubule allowed avoiding undesirable transverse oscillations of a particle in the course of its driving along a scaffold and it was directing.

The graph paper was glued to the outer side surface of the tube to determine the distance traveled by the particle at any one time. The study at 25 frames at maximum capacity until the end of the experiment was conducted . The data for the analysis for 1 second are used.

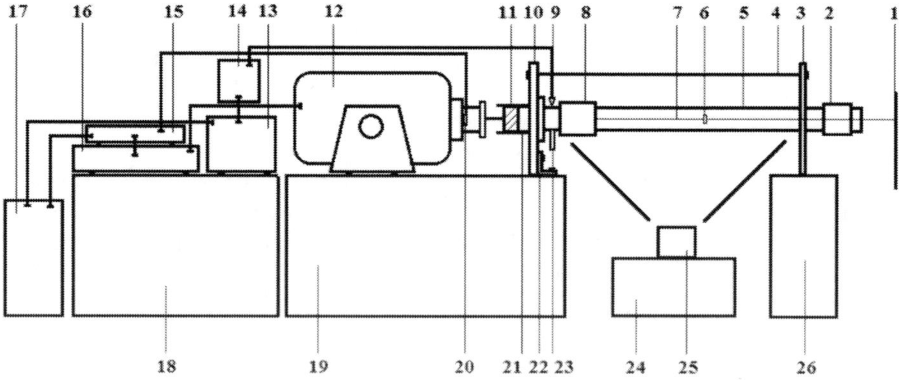

Figure 1. Schema of setup: 1 - stick, 2 - head at the open end, 3 - machine bed at the open end, 4 - plate, 5 – tube, 6 - particle 7 - fishing line, 8 - at the head of the piston, 9 - pressure sensor, 10 - machine bed in the piston, 11 - piston, 12 - vibration generator, 13 - oscilloscope, 14 - three-channel bridge amplifier voltage 15 - controller 16 – power, 17 – PC, 18 – table, 19 – stand, 20 - piezoelectric IEPE accelerometer, 21 - cylinder piston , 22- area, 23 - the rod with a needle, 24 - stand for camera, 25 - camera 26 - optional stand.

3. Experimental results

The dependence for the coordinates of the particles with m = 4.6 mg on the time along the tube at the resonance frequency ν_1 = 78.6 Hz gas column for different amplitudes of displacement of the piston l = 1.75 mm, l = 1.5 mm, l = 1.25 mm near the open end is shown in Figure 2. The initial coordinates of the particles were equal to 16.3 mm, 15.6 mm and 22.4 mm, respectively, for the convenience of data processing. We see that the particle begins to move from the open end of the tube of the piston. The behavior of the investigated particles due to the presence acoustic streaming in the form of a vortex motion of the gas in the tube. Her motion it is oscillatory, oscillates at a frequency of 4 Hz and an amplitude of 0.3 mm, the frequency is much smaller than the amplitude of the excitation of the gas, due to the friction of the particles on the fishing line. As we see, time of passing decreases with increase in amplitude of exaltation of gas. The distance traveled by the particle in amplitude l = 1.75 mm for 1 second, equal to 20.8 mm for the amplitudes l = 1.5 mm and l = 1.25 mm - 11.8 mm and 3.4 mm, respectively. The drift velocity of the particle matter under these amplitudes 20.8 mm/s, 11.8 mm/s and 3.4 mm/s. Increase the amplitude of excitation of the gas increases the amplitude of oscillations of the particle. The experiments were conducted when the particle is installed outside of the tube at a distance of 30 mm from the open end. It was not observed noticeable oscillations of a particle. A linear dependence was distances of time. The speed had values V = 520 mm/s and V = 250 mm/s, respectively. Increase of the amplitude the displacement of the piston increases the velocity of the particle.

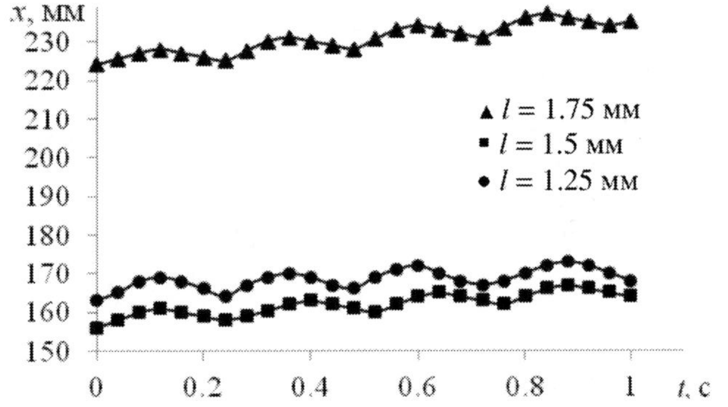

Figure 2. The dependence of coordinate is of a particle along a tube from time on a resonant frequency for various amplitudes of shifts of the piston (m = 4.6 mg, ν_1 = 78.6 Hz). Points – the experimental dates, solid lines - approximation.

The next experimental results are presented in which the particle is installed inside the tube immediately near an open end. The dependence for the coordinates of the particle with m = 4.6 mg axially from time to different oscillation frequencies with the amplitudes of the gas piston displacement l = 1.75 mm is represented in Figure 3. The mode is selected for the oscillations of the particle, when she did not commit a drift towards the inside and the outside of the tube. The Oscillations of the particles were of almost harmonic character of the experimental data. The increase in amplitude and oscillation frequency of a particle is observed with increase in an oscillation frequency of gas that is bound to increase in intensity of oscillations. Coordinate is displaced in the direction by the open end at which particle does not make a drift.

Acknowledgments
This work was funded by the subsidy of the Russian Government to support the Program of Competitive Growth of Kazan Federal University among World's Leading Academic Centers.

Journal of Physics: Conference Series **567** (2014) 012042 doi:10.1088/1742-6596/567/1/012042

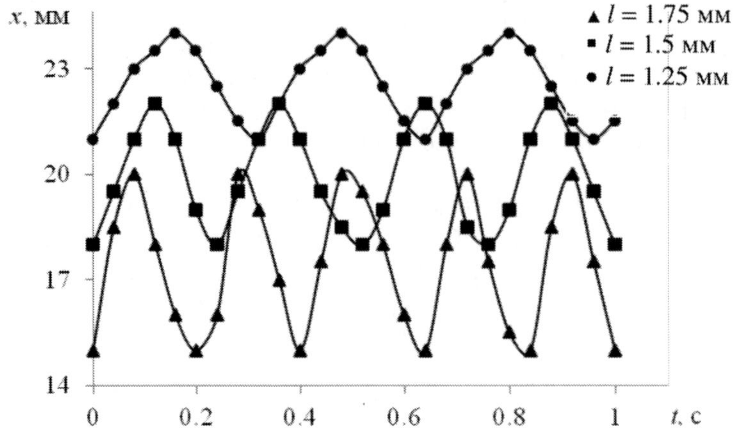

Figure 3. The ddependence of coordinate is of a particle with $m = 4.6$ mg along an axis in a tube near the open end from time on various oscillation frequencies of gas. Points – the experimental dates, solid lines - approximation.

References

1. Nigmatulin R I 1990 *Dynamics of Multiphase Media* (New York: Hemisphere)
2. Ganiev R F and Ukrainskii L E 1975 *Dynamics of Particles under the Effect of Vibration* (Kiev: Naukova Dumka) (in Russian)
3. Gubaidullin D A, Zaripov R G, Galiullin R G, Tkachenko L A 2007 *Journal of Problems of Power Engineering* (9-10) 3 (in Russian)
4. Gubaidullin D A, Zaripov R G, Galiullin R G, Tkachenko L A 2010 *Journal of Problems of Power Engineering* (9-10) 15 (in Russian)
5. Galiullin R G, Timokhina L A, Filipov S E 2001 *Acoustical Physics* **47**(5) 529
6. Ilgamov M A, Zaripov R G, Galiullin R G, Repin V B 1996 Applied Mechanics Reviews **49**(3) 137
7. Gubaidullin D A, Zaripov R G, Tkachenko L A 2012 Journal of Engineering Physics and Thermophysics **85**(2) 259

Institute of Physics
Dirac House, Temple Back
Bristol BS1 6BE UK

CURRAN ASSOCIATES INC
proceedings.com